本书英文版由世界卫生组织（World Health Organization）出版，书名为：

WHO TobLabNet Official Method SOP 11: Standard Operating Procedure for Determination of Nicotine, Glycerol and Propylene Glycol in E-liquids © World Health Organization 2021

WHO TobLabNet Official Method SOP 12: Standard Operating Procedure for Determination of Nicotine Content in Smokeless Tobacco Products © World Health Organization 2022

WHO TobLabNet Official Method SOP 13: Standard Operating Procedure for Determination of Moisture Content in Smokeless Tobacco Products © World Health Organization 2022

WHO TobLabNet Official Method SOP 14: Standard Operating Procedure for Determination of the pH of Smokeless Tobacco Products © World Health Organization 2022

WHO TobLabNet Official Method SOP 15: Standard Operating Procedure for Determination of Nicotine, Glycerol and Propylene Glycol Content in the Tobacco of Heated Tobacco Products © World Health Organization 2023

世界卫生组织（World Health Organization）授权中国科技出版传媒股份有限公司（科学出版社）翻译出版本书中文版。中文版的翻译质量和对原文的忠实性完全由科学出版社负责。当出现中文版与英文版不一致的情况时，应将英文版视作可靠和有约束力的版本。

中文版《世界卫生组织烟草实验室网络标准操作规程Ⅱ》

© 中国科技出版传媒股份有限公司（科学出版社） 2024

WHO TobLabNet SOP
世界卫生组织烟草实验室网络标准操作规程 II

胡清源　侯宏卫　主译

科学出版社

北京

内 容 简 介

本书为世界卫生组织（WHO）烟草实验室网络（TobLabNet）成员编写的标准操作规程（SOP）。包括电子烟烟液中烟碱、甘油和丙二醇含量的测定，无烟烟草制品中烟碱含量的测定，无烟烟草制品中水分含量的测定，无烟烟草制品 pH 的测定，加热型烟草制品中烟碱、甘油和丙二醇含量的测定共 5 个标准操作规程。

本书会引起吸烟与健康、烟草化学和公共卫生学等诸多领域研究人员的兴趣，可以为涉足烟草科学研究的科技工作者和烟草管制研究的决策者提供权威性参考。

图书在版编目（CIP）数据

世界卫生组织烟草实验室网络标准操作规程. Ⅱ / WHO 烟草实验室网络著；胡清源，侯宏卫主译. -- 北京：科学出版社，2024.6. -- ISBN 978-7-03-078993-8

Ⅰ. TS41-65

中国国家版本馆 CIP 数据核字第 2024XE2960 号

责任编辑：刘 冉 / 责任校对：杜子昂
责任印制：赵 博 / 封面设计：北京图阅盛世

科学出版社 出版
北京东黄城根北街 16 号
邮政编码：100717
http://www.sciencep.com

北京科印技术咨询服务有限公司数码印刷分部印刷
科学出版社发行　各地新华书店经销

*

2024 年 6 月第 一 版　开本：720×1000　1/16
2025 年 1 月第二次印刷　印张：8 1/2
字数：170 000

定价：120.00 元
（如有印装质量问题，我社负责调换）

译者名单

主　译：胡清源　侯宏卫
副主译：陈　欢　李　晓　崔利利
　　　　韩书磊　付亚宁
译　者：胡清源　侯宏卫　陈　欢
　　　　李　晓　崔利利　韩书磊
　　　　付亚宁　王红娟　田雨闪

目 录

WHO TobLabNet SOP 11 ·· 1
电子烟烟液中烟碱、甘油和丙二醇含量的测定标准操作规程··················· 1

WHO TobLabNet SOP 12 ·· 18
无烟烟草制品中烟碱含量的测定标准操作规程·· 18

WHO TobLabNet SOP 13 ·· 29
无烟烟草制品中水分含量的测定标准操作规程·· 29

WHO TobLabNet SOP 14 ·· 37
无烟烟草制品 pH 的测定标准操作规程·· 37

WHO TobLabNet SOP 15 ·· 44
加热型烟草制品中烟碱、甘油和丙二醇含量的测定标准操作规程··············· 44

CONTENTS

WHO TobLabNet SOP 11
Standard operating procedure for determination of nicotine, glycerol and propylene glycol in e-liquids 61

WHO TobLabNet SOP 12
Standard operating procedure for determination of nicotine content in smokeless tobacco products 81

WHO TobLabNet SOP 13
Standard operating procedure for determination of moisture content in smokeless tobacco products 94

WHO TobLabNet SOP 14
Standard operating procedure for determination of the pH of smokeless tobacco products 102

WHO TobLabNet SOP 15
Standard operating procedure for determination of nicotine, glycerol and propylene glycol content in the tobacco of heated tobacco products 109

WHO TobLabNet SOP 11

电子烟烟液中烟碱、甘油和丙二醇含量的测定

标准操作规程

方　　法：电子烟烟液中烟碱、甘油和丙二醇含量的测定
分析物：烟碱（3-[(2*S*)-1-甲基吡咯烷-2-基]吡啶）（CAS 号：54-11-5）
甘油（丙烷-1,2,3-三醇）（CAS 号：56-81-5）
丙二醇（丙烷-1,2-二醇）（CAS 号：57-55-6）
基　　质：电子烟烟液
更新时间：2021 年 3 月

本方法由世界卫生组织（WHO）烟草实验室网络（TobLabNet）成员实验室与欧洲烟草控制联合行动（JATC）成员实验室合作编写，作为电子烟烟液中烟碱、甘油和丙二醇含量的测定标准操作规程（SOP）。

引言

为了在全球范围内建立具有可比性的电子烟烟液的检测方法，需要电子烟烟液特定成分的一致性检测方法。2014年10月13~18日在俄罗斯莫斯科举行的世界卫生组织《烟草控制框架公约》（WHO FCTC）缔约方大会第六次会议（COP6）要求公约秘书处邀请世界卫生组织：①为缔约方大会第七次会议（COP7）编写一份关于电子烟碱传输系统（ENDS）和电子非烟碱传输系统（ENNDS）的专家报告，其中包括ENDS/ENNDS对健康影响的最新证据以及它们在戒烟方面的潜在作用和对烟草控制工作的影响；②评估政策选择，以实现WHO FCTC COP6(9)号决议第2段概述的目标；③考虑这些产品中特定成分和释放物的检测方法。

由于烟碱含量在世界上某些地区被限制在一定浓度（例如在欧盟，电子烟烟液中烟碱的最高浓度为20 mg/mL），因此烟碱被认为是电子烟烟液中要优先测量的成分。由于甘油和丙二醇是电子烟烟液的典型成分，并且可以与烟碱同时测量，因此这些成分也包含在SOP中。

本标准操作规程根据ISO 20714 [2.1]，描述了电子烟烟液中烟碱、甘油和丙二醇含量的测定方法。

1 适用范围

本方法适用于气相色谱法测定电子烟烟液中烟碱、甘油和丙二醇的含量。本方法的工作范围为烟碱 1~30 mg/mL，丙二醇 200~1000 mg/mL，甘油 200~1000 mg/mL。

2 参考标准

2.1 ISO 20714 (en)：电子烟烟液 测定电子烟碱传输系统烟液中的烟碱、丙二醇和甘油 气相色谱法（ISO 20714:2019，IDT）。

2.2 ISO 13276：烟草和烟草制品 烟碱纯度的测定 钨硅酸重量法。

2.3 ISO 5725-2：检测方法和结果的准确度（正确度和精密度） 第2部分 确定标准检测方法重复性和再现性的基本方法。

2.4 WHO TobLabNet SOP 02 烟草制品成分和释放物分析方法验证标准操作规程(世界卫生组织烟草实验室网络,日内瓦,2017年,https://www.who.int/tobacco/

publications/prod_regulation/standard-operation validation-02/en/，2020 年 12 月 10 日访问）。

2.5 联合国毒品和犯罪问题办公室（UNODC）《代表性药物取样指南》（维也纳，实验室和科学科，2009 年，http://www.unodc.org/documents/scientific/Drug_Sampling.pdf，2020 年 12 月 10 日访问）。

3 术语和定义

3.1 烟碱含量：电子烟烟液中的烟碱总量，以 mg/g 电子烟烟液表示。

3.2 电子烟烟液：用于雾化的含或不含烟碱且可用电子传输系统吸入的液体或凝胶。

3.3 电子烟碱传输系统/电子非烟碱传输系统：用于雾化吸入电子烟烟液的装置。

3.4 实验室样品：用于进行实验室检测的样品，由一次性或在规定时间内送到实验室的单一类型产品组成。

3.5 试样：从实验室样品中随机抽取的待测样品。所取样品数量应能代表实验室样品。

3.6 试料：从试样中随机抽取的用于单次测试的样品。所取样品数量应能代表试样。

3.7 预充式电子烟烟弹：可以直接连接到电子烟碱传输系统的预填充电子烟烟液的容器。

4 方法概述

4.1 将电子烟烟液置于室温后均质。

4.2 用异丙醇[7.6]和内标物[7.7][7.8]组成的稀释剂稀释电子烟烟液。

4.3 通过将分析物（烟碱、甘油、丙二醇）与相应内标物的色谱图峰面积比进行作图，建立已知浓度分析物的校准曲线，从而确定试料的分析物含量。

5 安全和环境预防措施

5.1 遵守所有化学实验活动的常规安全与环境预防措施。

5.2 使用本检测方法对某些产品进行检测和评估时所用材料或设备可能对环境有害。本检测方法无力解决所有与使用相关的安全问题。所有使用此方法的人都有责任咨询相关机构并遵循现有监管要求，制定健康与安全规程以及环境预防措施。

5.3 应特别注意避免吸入或皮肤接触危险化学品。在制备或处理未稀释材料、标准溶液、稀释液或收集样品时，应在通风橱中进行，并穿戴合适的实验服、手套和护目镜。

6 仪器和设备

常规实验仪器，特别是：

6.1 适用于重复移液 100 μL 的正位移移液器（如 HandyStep Electronic 或 Eppendorf Repeater stream）。

6.2 适用于重复移液 9.9 mL 的正位移移液器（如 HandyStep Electronic 或 Eppendorf Repeater stream）。

6.3 20 mL 透明平底样品瓶或其他合适的烧瓶。

6.4 用于固定烧瓶的3D旋转混合器（如 Stuart Scientific gyro rocker 或 MiuLab RH-18+ 3D Rotating Mixer）。

6.5 旋涡混合器。

6.6 配备氢火焰离子化检测器的气相色谱仪（GC-FID）。

6.7 能够分离溶剂、内标物、烟碱和其他成分色谱峰的毛细管气相色谱柱（如 Agilent DB-ALC1，30 m × 0.32 mm，1.8 μm）。

6.8 可容纳不同类型预充式电子烟烟弹的带螺旋盖的离心管，离心转速至少为 1500 r/min，相对离心力（RCF）≈315。

7 试剂和耗材

除非另有说明，所有试剂应至少为分析纯。试剂尽可能以 CAS 号标识。

7.1 载气：高纯度（>99.999%）的氦气（CAS 号：7440-59-7）。

7.2 辅助气体：用于氢火焰离子化检测器的高纯度（>99.999%）的空气和氢气（CAS 号：1333-74-0）。

7.3 纯度不低于98%的烟碱（CAS 号：54-11-5）。也可使用纯度不低于98%的水杨酸烟碱盐（CAS 号：29790-52-1）。

7.4 纯度不低于98%的甘油（CAS 号：56-81-5）。

7.5 纯度不低于98%的丙二醇（CAS 号：57-55-6）。

7.6 最大含水量为 1.0 g/L 的异丙醇（CAS 号：67-63-0）。

7.7 烟碱测定的气相色谱分析内标物：纯度不低于98%的正十七烷（CAS 号：629-78-7）、喹哪啶（CAS 号：91-63-4）或其他合适的替代品。

7.8 甘油和丙二醇测定的内标物：纯度不低于99%的1,3-丁二醇（CAS 号：107-88-0）。

8 玻璃器皿的准备

清洁、干燥的玻璃器皿，避免残留物污染。

9 溶液配制

9.1 稀释液。

含有适当浓度内标物(0.5 mL/L 正十七烷和 2 mL/L 1,3-丁二醇)的异丙醇[7.6]。将 0.50 mL 正十七烷[7.7]和 2.0 mL 1,3-丁二醇[7.8]移液至 1 L 容量瓶中。用异丙醇[7.6]稀释至刻度，混合均匀，并将溶液转移到具有防止污染功能的储存容器中。

注：考虑内标物对灵敏度和选择性的影响以及方法的线性范围，内标物的浓度和/或类型可以调整。

10 标准溶液配制

下述标准溶液配制方法仅供参考，如有需要可进行调整。

10.1 烟碱标准储备液（5 g/L）。

称取约 500 mg 烟碱[7.3]或 925 mg 水杨酸烟碱盐（精确至 0.1 mg），置于 100 mL 容量瓶中，用稀释液[9.1]稀释至刻度。

混合均匀，并在 0~4℃条件下避光储存（最长保存 4 周）。

在低温下储存的溶剂和溶液在使用前应平衡至(22 ± 5)℃。

10.2 甘油和丙二醇标准储备液（50 g/L）。

称取约 5000 mg 甘油和 5000 mg 丙二醇，精确至 0.1 mg，置于 100 mL 容量瓶中，用稀释液[9.1]稀释至刻度。

混合均匀，并在 0~4℃条件下避光储存（最长保存 4 周）。

在低温下储存的溶剂和溶液在使用前应平衡至(22 ± 5)℃。

10.3 标准溶液。

10.3.1 如表 1 所示，分别移取特定体积的烟碱标准储备液[10.1]至 100 mL 容量瓶中，并用稀释液[9.1]稀释至刻度。

10.3.2 如表 2 所示，分别移取特定体积的甘油和丙二醇标准储备液[10.2]至相同的 100 mL 容量瓶[10.3.1]中，并用稀释液[9.1]稀释至刻度。

在 4~8℃条件下避光储存（最长保存 1 周）。

10.3.3 标准溶液的烟碱、甘油和丙二醇最终浓度由下式确定：

$$最终浓度（mg/L）= \frac{x \cdot y}{10000}$$

式中：x——称取的分析物[10.1]或[10.2]的原始质量（mg）；

　　　y——移取的储备液[10.3.1]或[10.3.2]的体积（mL）。

标准溶液中烟碱的浓度如表1所示，甘油和丙二醇的浓度如表2所示。

表1　标准溶液中烟碱的浓度

标准溶液	5 g/L 烟碱储备液体积（mL）	内标液体积（μL）	总体积（mL）	标准溶液中烟碱最终浓度（mg/mL）	电子烟烟液中相应烟碱浓度（mg/mL）
1	0.2	不适用，包含在稀释液中	100	0.01	1.0
2	1.0		100	0.05	5.0
3	2.0		100	0.10	10.0
4	4.0		100	0.20	20.0
5	6.0		100	0.30	30.0

表2　标准溶液中甘油和丙二醇的浓度

标准溶液	50 g/L 甘油/丙二醇储备液体积（mL）	内标液体积（μL）	总体积（mL）	标准溶液中甘油/丙二醇最终浓度（mg/mL）	电子烟烟液中相应甘油/丙二醇浓度（mg/mL）
1	4.0	不适用，包含在稀释液中	100	2.0	200
2	8.0		100	4.0	400
3	12.0		100	6.0	600
4	16.0		100	8.0	800
5	20.0		100	10.0	1000

标准溶液范围可根据所使用的设备和待测样品进行调整，同时要考虑方法灵敏度可能对结果产生的影响。

在使用前，应将所有溶剂和溶液调节至室温(22 ± 5)℃。

11　抽样

11.1　样品收集。

根据适用的法规或国际标准，并考虑样品的可用性，选取具有代表性的电子烟烟液样品。

11.2　试样组成。

若可能，将实验室样品分成独立的单元（如包、条）。

对于每个试样，从实验室样品中抽取有代表性数量的产品。

每个试料应至少包含一个待测产品单元。

12 样品准备

电子烟烟液黏度较高，移液时必须特别小心，只能使用正位移移液器。内部实验室验证表明移液程序对结果有很大影响，因此在 SOP 中包含了专门的移液步骤。

如果没有可用的正位移移液器，可将电子烟烟液称量作为替代方法。然而，为了能够以 mg/mL 为单位报告结果，需要确定电子烟烟液的密度，并将电子烟烟液的质量换算为体积单位（mL）。

移取电子烟烟液的具体步骤在 **12.1** 中描述，从预充式烟弹中提取电子烟烟液的方法在 **12.2** 中描述。

12.1 加注瓶。

12.1.1 打开容器前,将电子烟烟液均质,避免气泡形成。如果有气泡，在气泡消失之前不要使用（电子烟烟液的超声将有助于去除气泡，但要避免超声引起的升温）。

12.1.2 吸取 1 mL 不含气泡的电子烟烟液，可将移液器设为 100 μL，重复吸取 10 次。

12.1.3 用枪头外侧碰触样品容器边缘，去除多余的液体。

12.1.4 吸弃 2 次 100 μL 至废液桶，枪头不要碰触废液桶。

12.1.5 检查枪头外部是否有电子烟液残留，如有，重复步骤 **12.1.3**。

12.1.6 再一次吸弃 2 次 100 μL 至废液桶，枪头不要碰触废液桶。

12.1.7 检查枪头外部是否有电子烟烟液残留。

12.1.8 吸取 100 μL 电子烟液（$V_{\text{e-liquid}}$）至 20 mL 样品瓶[6.3]。

12.1.9 将剩余的电子烟烟液吸弃至废液桶。

12.1.10 尽快关上电子烟烟液容器，防止样品蒸发或污染。

12.2 预充式烟弹。

为了能够分析预充式烟弹中的电子烟烟液，需要从烟弹中提取烟液。为了避免电子烟烟液被烟弹的组件或打开烟弹的工具污染，下面描述了使用离心机从烟弹中提取总量的电子烟烟液的步骤。

考虑到从烟弹中提取的电子烟烟液须是烟弹的代表性样品，也可使用替代方法。

12.2.1 将一次性枪头（100~1000 μL）剪下一块，放入合适的离心管

[6.8]中。

- **12.2.2** 取下电子烟烟弹的上盖。
- **12.2.3** 将烟弹顶部朝下放入离心管中，置于枪头上方。
- **12.2.4** 使用螺旋盖盖上离心管。
- **12.2.5** 离心管在 1500 r/min 转速下（相对离心力 RCF≈315）离心至少 5 min。
- **12.2.6** 将烟弹和枪头从离心管中移除（必要时使用镊子，避免污染）。
- **12.2.7** 按照 **12.1** 的方法对烟液进行移液。

注：当用于移液的电子烟烟液少于 1 mL 时，需用另一种移液方法。

例如，如果只有 600 μL 电子烟烟液，那么吸取 400 μL（而不是 1 mL），并在第一次吸弃至废液桶之后将 100 μL 电子烟烟液移液至样品瓶中。

如果可用的电子烟烟液少于 500 μL，则尽可能多地吸取电子烟烟液，并将 100 μL 电子烟烟液移液到样品瓶中，而不将电子烟烟液吸弃至废液桶中。如果可用的电子烟烟液少于 100 μL，则移液量尽可能高。

如果使用不同的方法（可用量小于 1 mL），请在分析报告中详细说明移液步骤。

13　吸烟机设置

不适用。

14　样品生成

不适用。

15　样品制备

15.1 用正位移移液器[6.2]将 9.9 mL 稀释液[9.1]加入到含有 100 μL 电子烟烟液的 20 mL 样品瓶中，并关上样品瓶。

15.2 使用涡旋混合器[6.5]将样品瓶中的液体均质化至少 30 s。设置旋涡混合器的速度，使液体均质化过程中产生涡流。

15.3 将均质液体（测试溶液）的试料转移到与 GC-FID 仪器兼容的自动进样器样品瓶中。

15.4 如果要保存样品，请在 4~8℃避光保存（最长保存 1 周）。

注：确保均质后液体没有视觉上的不均匀。如果仍然可以看到不均匀性，则需根据 **15.2** 重复进行均质化，直到没有视觉上的不均匀。

16 样品分析

使用配备氢火焰离子化检测器的气相色谱仪测定电子烟烟液中的烟碱、甘油和丙二醇。分析物可在所用的色谱柱上与其他潜在干扰物分离开。通过比较测试溶液的峰面积比与已知浓度标准溶液的峰面积比（分析物与内标物），得到分析物的浓度。

16.1 GC 操作条件示例。

GC 柱：Agilent DB-ALC1（30 m × 0.32 mm，1.8 μm）或同等产品。

柱程序：

初始温度：	140℃；
保持时间：	5 min；
温度速率：	40℃/min；
最终温度：	250℃；
保持时间：	4 min；
进样口温度：	225℃；
检测器温度：	260℃；
载气：	流速为 1.5 mL/min 的氦气；
进样体积：	1.0 μL；
进样方式：	分流 1：50。

注：根据仪器和色谱柱条件以及色谱峰的分离度调整操作参数。

附录 2 提供了使用气相色谱-质谱（GC-MS）作为 GC-FID 的替代方法的设置示例。

16.2 预计保留时间。

对于此处描述的条件，预期的洗脱顺序是丙二醇、1,3-丁二醇、甘油、烟碱和正十七烷。

注 1：温度、气体流速和色谱柱的使用年限的差异可能会改变保留时间。

注 2：在分析开始之前，必须验证洗脱顺序和保留时间。

注 3：在上述条件下，预计总分析时间约为 11 min（延长分析时间可以优化性能）。

16.3 烟碱、甘油和丙二醇的测定。

测试样品的测量顺序应按照适用的实验室质量体系设计。本节给出一个确定电子烟烟液中烟碱、甘油和丙二醇的测定顺序的示例。

在相同的条件下注入等量的标准溶液和测试溶液。

16.3.1 在使用前通过注入两份 1 μL 的样品溶液作为预试样来调节系统。

16.3.2 在与样品相同的条件下注入 1 μL 稀释液[9.1]和一份标准溶液，验证气相色谱系统的性能和所用试剂的成分污染情况。

16.3.3 将各标准溶液（烟碱、甘油和丙二醇标准溶液）注入气相色谱仪。

16.3.4 评估标准溶液的保留时间和响应（面积计数）。如果保留时间与之前进样的保留时间相似（±0.2 min），并且响应值与之前进样的典型响应值的差异在20%以内，则系统已准备就绪，可进行分析。如果保留时间和/或响应值超出规范范围，则应根据各实验室政策采取纠正措施。

16.3.5 记录烟碱、甘油、丙二醇及内标物峰面积。

16.3.6 计算标准溶液（包括空白溶剂）中每种分析物的分析物峰与内标峰的相对响应比（RF = $A_{analyte}/A_{IS}$）。

16.3.7 绘制相对响应比（Y轴）与分析物浓度（X轴）的曲线图。

16.3.8 标准曲线应在整个范围内呈线性。

16.3.9 使用线性回归的斜率（b）和截距（a）来计算线性回归方程（$Y = a + bx$）。如果 $R^2<0.99$，则应重新校准。根据实验室程序检查个别异常值。

16.3.10 注入 1 μL 的质量控制样品和试料，用相应软件测定峰面积。

16.3.11 所有测试溶液获得的分析物信号（峰面积）必须在校准曲线的工作范围内；否则，溶液应根据需要用稀释液[9.1]稀释。

注：代表性色谱图见附录1。

17 数据分析与计算

17.1 对于每个试料，计算分析物峰面积与内标峰面积的比值（Y_t）。

17.2 使用线性回归系数计算每个试料中分析物浓度（C_{TP}，mg/mL）：

$$C_{TP} = (Y_t - a)/b$$

17.3 根据下式计算以 mg/mL 表示的电子烟烟液样品中分析物浓度（$C_{e\text{-liquid}}$）：

$$C_{e\text{-liquid}} = C_{TP} \times (V_{tot}/V_{e\text{-liquid}})$$

式中：$C_{e\text{-liquid}}$——电子烟烟液中分析物的浓度（mg/mL）；

C_{TP}——试料中分析物的浓度（mg/mL）；

V_{tot}——电子烟烟液稀释后试料的总体积（mL，默认值为 10 mL）；

$V_{e\text{-liquid}}$——在 12.1.8 中使用的电子烟烟液的体积（mL）。

18 特别注意事项

18.1 安装新色谱柱后,按所述 GC 条件注入电子烟烟液样品溶液。应重复进样,直到该组分和内标物的峰面积(或峰高)可重现。这需要大约 4 次进样。

18.2 每组(序列)样品结束后,建议将色谱柱温度升高至 220℃ 30 min,从气相色谱柱中清除高沸点成分。

18.3 当内标物的峰面积(或峰高)明显高于预期时,建议用不加内标物的稀释液稀释电子烟烟液样品,并按本操作规程进行分析。这样就可以确定是否有其他成分与内标物共洗脱,从而导致分析物数值偏低的误差。

19 数据报告

19.1 报告每个评估样品的单独测量值。
19.2 按照方法规范的要求报告结果。
注:更多信息可参见 WHO TobLabNet SOP 02[2.4]。

20 质量控制

20.1 控制参数。
注 1:如果控制测量值超出预期值的公差极限,则必须根据实验室质量程序进行适当的调查并采取措施。
注 2:如有必要,需要做额外的实验室质量保证规程,以符合各个实验室的做法。
20.2 实验室试剂空白样。
如 **16.3.2** 所述,为了在样品制备和分析过程中检测潜在的污染,需包含稀释溶液的空白测定[**9.1**]。结果应小于相应分析物的检出限。
20.3 质量控制样品。
为了验证整个分析的一致性,根据各实验室的实践分析参比或质量控制电子烟烟液。

21 方法性能

21.1 检测限(LOD)和定量限(LOQ)。
LOD 定义为试样中被测物能被检测出的最低水平,通过已知低浓度的被测物的检测信号与基线噪声比较,确定可被检测出的最小量或浓度。检测限以信噪比为3∶1来界定。LOQ设置为LOD的两倍。如表3所示（数据由非内部实验室验证提供）。

表3 电子烟烟液中烟碱、甘油和丙二醇的LOD和LOQ

成分	LOD (mg/mL)	LOQ (mg/mL)
烟碱	0.1	0.2
甘油	1.0	2.0
丙二醇	0.5	1.0

21.2 实验室基质加标回收率。

加入到基质上的分析物的回收率用作准确度的替代量度。这些成分的回收率是通过内部实验室验证和国际合作研究确定的。

在内部实验室验证中，通过将不同量的烟碱、甘油和丙二醇称量到1 L烧瓶中制备回收样品来确定回收率。使用3D旋转混合器[6.4]将回收样品均质2 h后，将回收样品转移到2 mL的小瓶中。在每个回收样品中测定烟碱、甘油和丙二醇，在一天内对单个小瓶进行五次分析。未加标的样品也要分析。回收率由下式计算，烟碱的回收率如表4所示，甘油和丙二醇的回收率如表5所示。

回收率(%) = 100 × (分析结果－未加标结果)/加标量

表4 通过内部实验室验证的烟碱的平均值和回收率

加标量（mg/mL）	烟碱	
	平均值（mg/mL）	回收率（%）
0.195	0.235	120.6
3.098	3.117	96.5
12.388	12.410	101.1
20.209	20.232	98.9

表5 通过内部实验室验证的甘油和丙二醇的平均值和回收率

甘油			丙二醇		
加标量（mg/mL）	平均值（mg/mL）	回收率（%）	加标量（mg/mL）	平均值（mg/mL）	回收率（%）
378.7	377.7	99.7	727.4	723.1	99.4
630.5	636.5	101.0	519.6	515.1	99.1
883.0	898.2	101.7	311.2	307.3	98.7

在一项国际合作研究中，通过将4种不同量的烟碱、甘油和丙二醇称量到烧瓶中，确定了回收率。使用3D旋转混合器[6.4]将烧瓶匀浆后，将回收样品转移到10 mL烧瓶中。在一项2020年进行的国际合作研究中，通过对单个烧瓶进行两次分析，测定了4种加标样品中的烟碱、甘油和丙二醇。回收率由下式计算，烟碱

的回收率见表 6，甘油和丙二醇的回收率见表 7。

回收率(%) = 100 × (分析结果 – 未加标结果)/加标量

表 6　国际合作研究中获得的烟碱的平均值和回收率

样品	烟碱		
	理论浓度（mg/mL）	平均值（mg/mL）	回收率（%）
A	0.25	0.326	130.4
B	5.08	5.08	100.0
C	8.03	7.91	98.5
D	21.27	22.06	103.7

表 7　国际合作研究中获得的甘油和丙二醇的平均值和回收率

样本	甘油（mg/mL）			丙二醇（mg/mL）		
	理论浓度（mg/mL）	平均值（mg/mL）	回收率（%）	理论浓度（mg/mL）	平均值（mg/mL）	回收率（%）
A	568.0	567.5	99.9	568.7	563.1	99.0
B	213.9	211.5	98.9	855.0	843.2	98.6
C	772.5	738.0	95.5	278.3	268.4	96.4
D	321.5	316.7	98.5	749.9	736.8	98.2

21.3　分析特异性。

目标分析物的保留时间用于验证分析特异性。分析物与质量控制电子烟烟液内部标准响应比的范围用于验证未知样品结果的特异性。

21.4　线性。

建立的烟碱校准曲线在 0.01~0.30 mg/mL（1.0~30 mg/mL）标准浓度范围内呈线性。

丙二醇和甘油在 2.0~10.0 mg/mL（200~1000 mg/mL）标准浓度范围内呈线性。

21.5　潜在干扰。

由于与其中一种成分或内部标准成分的保留时间相似，调味物质的存在会引起干扰。

22　重复性和再现性

2020 年进行的一项国际合作研究[23.4]涉及 23 个实验室和 5 个样品（4 个加标电子烟烟液和 1 个商品电子烟烟液），给出了本方法的以下值。

常规正确操作该方法时，同一操作者使用同一设备在最短的操作时间内匹配的电子烟烟液样品的两个单一结果之间的差异将超过重复性限 r 的情况，平均 20 个样品不超过一次。

常规正确操作该方法时,两个实验室报告的匹配电子烟烟液样品的单一结果差异超过再现性限 R 的情况,平均 20 次不超过一次。

根据 ISO 5725-1[23.1]和 ISO 5725-2[2.3]对测试结果进行统计分析,得出表 8 至表 10 所示的准确数据。

表 8 电子烟烟液中烟碱（mg/mL）测定的精密度限值

电子烟烟液	n	$\hat{m}$	r	R
样品 A	15	0.326	0.079	0.277
样品 B	22	5.08	0.52	1.06
样品 C	21	7.91	0.26	1.49
样品 D	21	22.06	0.66	3.74
样品 E	22	11.38	1.30	1.93

注：n——参与的实验室数量；
$\hat{m}$——电子烟烟液中烟碱含量的平均值；
r——电子烟烟液中烟碱含量的重复性限；
R——电子烟烟液中烟碱含量的再现性限。

表 9 电子烟烟液中甘油（mg/mL）测定的精密度限值

电子烟烟液	n	$\hat{m}$	r	R
样品 A	20	567.5	13.5	48.1
样品 B	21	211.5	8.5	43.6
样品 C	21	738.0	49.3	69.9
样品 D	20	316.7	17.2	32.1
样品 E	21	365.5	16.1	40.7

注：n——参与的实验室数量；
$\hat{m}$——电子烟烟液中甘油含量的平均值；
r——电子烟烟液中甘油含量的重复性限；
R——电子烟烟液中甘油含量的再现性限。

表 10 电子烟烟液中丙二醇（mg/mL）测定的精密度限值

电子烟烟液	n	$\hat{m}$	r	R
样品 A	22	563.1	13.1	23.2
样品 B	22	843.2	37.6	60.7
样品 C	22	268.4	15.5	18.3
样品 D	22	736.8	21.5	38.6
样品 E	23	570.1	58.3	58.3

注：n——参与的实验室数量；
$\hat{m}$——电子烟烟液中丙二醇含量的平均值；
r——电子烟烟液中丙二醇含量的重复性限；
R——电子烟烟液中丙二醇含量的再现性限。

23 参考文献

23.1 ISO 5725-1：检测方法和结果的准确度（正确度和精密度） 第1部分 一般原则和定义。

23.2 ISO 5725-4：检测方法和结果的准确度（正确度和精密度） 第4部分 确定标准检测方法正确度的基本方法。

23.3 ISO/TC 126 (http://www.iso.org/iso/home/store/cataloge_tc/cataloge_tc_browser.tm?commid=52158)。

23.4 电子烟烟液中测定烟碱、甘油和丙二醇的分析方法验证的合作研究报告（待出版）。

附录1 电子烟烟液中烟碱、丙二醇和甘油含量测定的典型色谱图

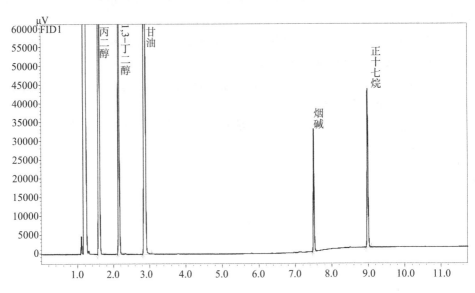

附图1 烟碱浓度为0.30 mg/mL、丙二醇和甘油浓度为10.0 mg/mL的标准溶液色谱图示例

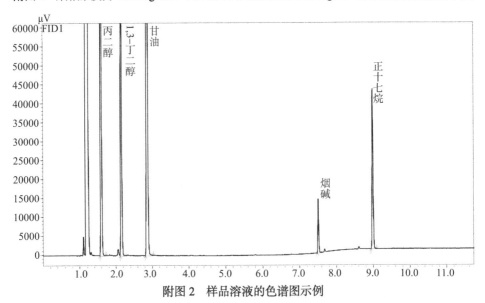

附图2 样品溶液的色谱图示例

附录2　GC-MS 作为替代测量方法的设置示例

特定的 GC 条件,如载气、流速和总运行时间,可根据特定的 MS 需求调整。
MS 操作条件:
传输线温度:≥180℃;
停留时间:50 ms;
电离模式:电子电离(电离电压 70 eV)。
检测:
丙二醇:m/z 61(定量离子),45(定性离子);
甘油:m/z 61(定量离子),43(定性离子);
烟碱:m/z 162(定量离子),133(定性离子);
喹哪啶:m/z 143(定量离子),128(定性离子);
正十七烷:m/z 240(定量离子),85(定性离子),由于其比质谱低,不建议使用本内标进行 GC-MS 检测。

WHO TobLabNet SOP 12

无烟烟草制品中烟碱含量的测定标准操作规程

方　　法：无烟烟草制品中烟碱含量的测定
分　析　物：烟碱（3-[(2*S*)-1-甲基吡咯烷基-2-基]吡啶）（CAS 号：54-11-5）
基　　质：无烟烟草制品
更新时间：2021 年 12 月

无烟烟草制品正逐渐引起公共卫生组织的关注。2012年在首尔举行的世界卫生组织《烟草控制框架公约》（WHO FCTC）缔约方大会第五次会议（COP5）要求确定管制无烟烟草制品中的化学成分的备选办法。本文件是应2016年德里举行的缔约方大会第七次会议（COP7）向WHO FCTC秘书处提出的邀请世界卫生组织根据FCTC/COP6(12) 2b.ii号决议中最终确定测量烟碱和烟草特定亚硝胺的标准操作规程的要求编写的。根据这一要求，WHO组织了一项涉及其烟草实验室网络（TobLabNet）测试实验室的合作研究，测试了具有某些化学特性的材料，这些材料代表了一系列常见形式的无烟烟草制品，并且具有不同的物理和化学特性。本标准操作规程对经验证的WHO标准操作规程对无烟烟草制品的适用性进行了评价，并提出了推荐的方法。

本方法由世界卫生组织（WHO）烟草实验室网络（TobLabNet）成员实验室编写，作为无烟烟草制品中烟碱含量测定的标准操作规程（SOP）。

引言

为了在全球范围内建立具有可比性的烟草制品检测方法，需要特定参数的一致性检测方法。WHO TobLabNet审查了测定无烟烟草制品中烟碱的常用操作，以便编制标准操作规程作为WHO TobLabNet SOP。

本标准操作规程改编自WHO TobLabNet SOP 04[2.1]，描述了无烟烟草制品中烟碱测定的方法。

1 适用范围

本方法适用于气相色谱-氢火焰离子化检测法测定无烟烟草制品中烟碱的含量。

2 参考标准

2.1 WHO TobLabNet SOP 04 卷烟烟丝中烟碱的测定标准操作规程（世界卫生组织烟草实验室网络，日内瓦，2017年）。

2.2 联合国毒品和犯罪问题办事处（UNODC）《代表性药品抽样指南》（维也纳，实验室和科学科，2009年，http://www.unodc.org/documents/scientific/Drug_Sampling.pdf）。

2.3 WHO TobLabNet SOP 02 烟草制品成分和释放物分析方法验证标准操作规程（世界卫生组织烟草实验室网络，日内瓦，2017年）。

2.4 ISO 5725-1：检测方法和结果的准确度（正确度和精密度） 第1部分

一般原则和定义。

2.5 ISO 5725-2：检测方法和结果的准确度（正确度和精密度） 第 2 部分 确定标准检测方法重复性和再现性的基本方法。

3 术语和定义

3.1 烟碱含量：无烟烟草制品中的烟碱总量，以 mg/g 表示。

3.2 无烟烟草：无烟烟草制品中含有烟草的部分。

3.3 无烟烟草制品：完全或部分以烟叶为原料制成的产品，用于吸吮、咀嚼或鼻吸（世界卫生组织《烟草控制框架公约》第 1(f)条），包括鼻烟（干鼻烟和湿鼻烟）、咀嚼烟草或源自烟草植物的混合材料。

3.4 实验室样品：用于实验室检测的样品，由一次性或在规定时间内送到实验室的单一类型产品组成。

3.5 试样：从实验室样品中随机抽取的待测样品。所取样品数量应能代表实验室样品。

3.6 试料：从试样中随机抽取的用于单次测试的样品。所取样品数量应能代表试样。

4 方法概述

4.1 用正己烷、氢氧化钠溶液和水的混合物从无烟烟草中萃取烟碱。

4.2 采用气相色谱氢火焰离子化检测器对有机层进行分析。

4.3 通过分析物与内标物的色谱图峰面积比进行作图，建立已知浓度烟碱的校准曲线，从而确定试料的烟碱含量。

5 安全和环境预防措施

5.1 遵守所有化学实验室活动的常规安全与环境预防措施。

5.2 使用本检测方法对某些产品进行检测和评估时所用材料或设备可能对环境有害。本检测方法无力解决所有与使用相关的安全问题。所有使用此方法的人都有责任咨询相关机构并遵循现有监管要求，制定健康与安全规程以及环境预防措施。

5.3 应特别注意避免吸入或皮肤接触危险化学品。在制备或处理未稀释材料、标准溶液、稀释液或收集样品时，应在通风橱中进行，并穿戴合适的实验服、手套和护目镜。

6 仪器和设备

常规实验仪器，特别是：

6.1 萃取瓶：带有塞子的锥形瓶（100 mL），带有波纹密封盖和隔层的隔热玻璃瓶（100 mL），带有聚四氟乙烯塞子的培养管或其他合适的烧瓶。

6.2 使萃取瓶保持在适当位置的线性振荡器。

6.3 配备氢火焰离子化检测器的毛细管气相色谱仪。

6.4 能够明显分离溶剂、内标物、烟碱和其他烟草成分色谱峰的毛细管气相色谱柱（如 Varian WCOT 熔融石英色谱柱，25 m × 0.25 mm ID；涂层：CP-WAX 51）。

6.5 超声波水浴。

7 试剂和耗材

除非另有说明，所有试剂应至少为分析纯。试剂尽可能以 CAS 号标识。

7.1 载气：高纯度（>99.999%）氦气（CAS 号：7440-59-7）。

7.2 辅助气体：用于氢火焰离子化检测器的高纯度（>99.999%）的空气和氢气（CAS 号：1333-74-0）。

7.3 最大含水量为 1.0 g/L 的正己烷（CAS 号：110-54-3）（色谱纯）。

7.4 纯度不低于 98%的烟碱（CAS 号：54-11-5）。也可使用纯度不低于 98%的水杨酸烟碱盐（CAS 号：29790-52-1）。

7.5 氢氧化钠（CAS 号：1310-73-2）颗粒。

7.6 内标物：正十七烷（纯度不低于98%，质量分数）（CAS 号：629-78-7）。可使用喹哪啶（CAS 号：91-63-4）、异喹啉（CAS 号：119-65-3）、喹啉（CAS 号：91-22-5）或其他合适的替代品。

8 玻璃器皿的准备

清洁、干燥的玻璃器皿，避免残留物污染。

9 溶液配制

9.1 氢氧化钠溶液（2 mol /L）。

    9.1.1 称取约 80 g 氢氧化钠。

    9.1.2 将称好的氢氧化钠溶于水中，并用水稀释至 1 L。

9.2 萃取液（0.5 mg/mL）。

  9.2.1 称取约 0.5 g（精确至 0.001 g）正十七烷或其他可替代物作内标物。

  9.2.2 将称好的正十七烷或替代内标物溶于正己烷中并用正己烷稀释至 1 L。

10 标准溶液配制

下述标准溶液配制方法仅供参考，如有需要可进行调整。

10.1 烟碱标准储备液（2 g/L）。

  10.1.1 称取约 200 mg 烟碱或 370 mg 水杨酸烟碱盐（精确至 0.0001 g），置于 200 mL（或 250 mL）锥形瓶中。

  10.1.2 将称好的烟碱溶解于 50 mL 水中。

  10.1.3 移取 100 mL 萃取液[9.2.2]，并加入 25 mL 2 mol/L 氢氧化钠溶液。

  10.1.4 将获得的双相混合物在振荡器中剧烈振荡(60 ± 2) min。混合均匀。

  10.1.5 分离上层有机相,配制标准溶液。如有必要,在 4~8℃ 避光储存。

10.2 烟碱标准溶液。

  10.2.1 用 **10.1.4** 配制的标准储备液配制标准溶液，方案如表 1 所示。

  10.2.2 在容量瓶中加入萃取液[9.2.2]至刻度。

  10.2.3 在 4~8℃ 下避光储存标准溶液。

  10.2.4 确定标准溶液中烟碱的最终浓度：

$$最终浓度（mg/L）= \frac{x \times y \times 1000}{100 \times 20}$$

式中：x——烟碱[10.1.1]的原始质量（mg）；

    y——移取的储备液[10.2.1]的体积。

标准溶液中烟碱的浓度如表 1 所示。

表 1 标准溶液中烟碱的浓度

标准溶液	2 g/L 烟碱储备液体积（mL）	内标液体积（μL）	总体积（mL）	标准溶液中烟碱最终浓度（mg/L）	在称取 1.5 g 样品时，相当于无烟烟草中的近似水平（mg/g）
1	0.5	不适用，包含在萃取液中	20	50	1.3
2	2.5		20	250	6.7
3	5.0		20	500	13.3
4	7.5		20	750	20.0
5	10.0		20	1000	26.7
6	15.0		20	1500	40.0

标准溶液的范围可根据使用的设备和待测样品进行调整，注意可能对方法灵敏度产生的影响。

在使用前，应将所有溶剂和溶液调节至室温。

11 抽样

11.1 按照实验室抽样程序对无烟烟草制品进行抽样。可根据各实验室惯例或者要求的特殊规定或样品可用性，使用替代方法来获取具有代表性的实验室样品。

11.2 试样组成。

11.2.1 若可能，将实验室样品分成独立的单元（如包、条）。

11.2.2 从至少 $\sqrt{n}$ 个单元[2.2]中取出每种试样的等量产品。

12 样品准备

12.1 将无烟烟草制品从包或条中取出。包括质量控制样品（如适用）。

12.2 根据个别实验室惯例（例如可采用食品分析抽样方法），从无烟烟草制品中选取适量且具有代表性的部分。

12.3 从无烟烟草制品中提取无烟烟草。

12.4 将足够的无烟烟草样品混合，使每个试料约含 0.5~2 g 均质无烟烟草。

13 吸烟机设置

不适用。

14 样品生成

不适用。

15 样品制备

15.1 称取 0.5~2 g 试料，精确至 0.001 g，置于 100 mL 萃取容器中。

15.2 将试料与 20 mL 水、40 mL 萃取液[**9.2.2**]和 10 mL 2 mol/L 氢氧化钠溶液混合。

15.3 在振荡器上振动烧瓶(60 ± 2) min。

15.4 将样品瓶静置 20 min 后，使两相澄清分离。分离各相，用气相色谱尽快分析有机相（上层）。为了便于分离各相，可将锥形瓶置于超声波水浴中。

15.5 如果发现溶液浑浊，需进行另外的步骤：

（选项 1）用 Whatman 41 号纸过滤样品。

（选项 2）以 10000 r/min 或合适的相对离心力（RCF > 500）对样品进行离心，以确保获得澄清溶液。

15.6 如果要保存萃取的样品，于 4~8℃ 避光保存。

16 样品分析

使用配备氢火焰离子化检测器的气相色谱仪测定无烟烟草制品中的烟碱。分析物可在所用的色谱柱上与其他潜在干扰物分离开。通过比较测试溶液的峰面积比与已知浓度标准溶液的峰面积比（分析物与内标物），得到分析物的浓度。

16.1 GC 操作条件示例。

GC 柱：Varian WCOT 熔融石英毛细管柱，25 m × 0.25 mm ID；

涂层：CP-WAX 51 或替代物；

色谱柱温度：170℃（恒温）；

进样口温度：270℃；

检测器温度：270℃；

载气：流速为 1.5 mL/min 的氦气；

进样体积：1.0 μL；

进样方式：分流 1∶10。

注：根据仪器和色谱柱条件以及色谱峰的分离度调整操作参数。

16.2 预计保留时间。

  16.2.1 对于此处描述的条件，预期的洗脱顺序是正十七烷、烟碱。

  16.2.2 温度、气体流速和色谱柱的使用年限的差异可能会改变保留时间。

  16.2.3 在分析开始之前，必须验证洗脱顺序和保留时间。

  16.2.4 在上述条件下，预计总分析时间约为 6 min（延长分析时间可以优化性能）。

16.3 烟碱的测定。

根据各实验室的惯例来确定测试样品的测量顺序。本节给出一个确定无烟烟草制品中烟碱测定顺序的示例。

  16.3.1 注入正己烷[7.3]，检查系统或试剂中是否存在污染。

  16.3.2 在使用前通过注入两份 1 μL 的样品溶液作为预试样来调节系统。

  16.3.3 在与样品相同的条件下，注入 1 μL 萃取液[9.2]和一份标准溶液，验证气相色谱系统的性能。

16.3.4 注入 1 μL 正己烷[7.3]，检查系统或试剂是否有污染。

16.3.5 将每种烟碱标准溶液按随机顺序注入气相色谱仪。

16.3.6 评估标准溶液的保留时间和响应（面积计数）。如果保留时间与之前进样的保留时间相似（±0.2 min），并且响应值与之前进样的典型响应值的差异在 20%以内，则系统已准备就绪，可进行分析。如果保留时间和/或响应值超出规范范围，则应根据各实验室政策采取纠正措施。

16.3.7 记录烟碱和内标物峰面积。

16.3.8 计算每种烟碱标准溶液（包括溶剂空白溶液）烟碱峰与内标峰的相对响应比（RF = $A_{\text{nicotine}}/A_{\text{IS}}$）。

16.3.9 以烟碱浓度（X 轴）与峰面积比（Y 轴）绘制标准曲线图。

16.3.10 截距在统计上不应与零有显著差异。

16.3.11 标准曲线在整个标准范围内呈线性。

16.3.12 使用线性回归的斜率（b）和截距（a）来计算线性回归方程（$Y = a + bx$），若 $R^2 < 0.99$，则应重新校准。如果单个校准点与预期值（通过线性回归估计）相差超过 10%，则应舍弃该点。

16.3.13 注入 1 μL 的质量控制样品[20.3]和试料[15.4]，用相应软件测定峰面积。

16.3.14 所有试料的信号（峰面积）必须落在标准曲线的工作范围内；否则，应调整标准溶液或试料的浓度。

注：代表性色谱图见附录1。

17 数据分析与计算

17.1 对于每个试料，计算烟碱峰面积与内标峰面积的比值（Y_t）。

17.2 使用线性回归系数计算每个试料等分试样的烟碱浓度（mg/L）：

$$m_t = (Y_t - a)/b。$$

17.3 根据下式计算以 mg/g 表示的烟草样品中的烟碱含量（m_n）：

$$m_n = \frac{m_t \times V_e}{m_o \times 1000}$$

式中：m_t——试样溶液中烟碱的浓度（mg/L）；

V_e——萃取液的体积（mL）；

m_o——试料的质量（g）。

18 特别注意事项

18.1 安装新色谱柱后,按所述 GC 条件注入烟草样品萃取物对其进行调节。应重复进样,直到烟碱和内标物的峰面积(或峰高)可重现。这需要大约 4 次进样。

18.2 每组(序列)样品结束后,建议将色谱柱温度升高至 220℃ 30 min,从气相色谱柱中清除高沸点成分。

18.3 当内标物的峰面积(或峰高)明显高于预期时,建议用不含内标物的萃取液萃取烟草样品。这样就可以确定是否有其他组分与内标物共洗脱,从而导致烟碱值偏低的误差。

19 数据报告

19.1 报告每个评估样品的单独测量值。
19.2 按照方法规范的要求报告结果。
19.3 更多信息可参见 WHO TobLabNet SOP 02[2.3]。

20 质量控制

20.1 控制参数。
注 1：如果控制测量值超出预期值的公差极限,则必须进行适当的调查并采取措施。
注 2：如有必要,需要做额外的实验室质量保证规程,以符合各个实验室的做法。
20.2 实验室试剂空白样。
如 16.3.4 所述,为了在样品制备和分析过程中检测潜在的污染,需包含实验室试剂空白样。空白样由分析测试样品中使用的所有试剂和材料组成,并像测试样品一样进行分析。结果应小于相应分析物的检出限。
20.3 质量控制样品。
为了验证整个分析的一致性,根据各实验室的实践分析参比烟草制品,如 CORESTA 参比产品(CRP)。

21 方法性能

21.1 报告限。
报告限设置为所用的标准曲线的最低浓度,重新计算为 mg/g(例如,1.3 mg/g 对应 50 mg/L 最低标准浓度)。

21.2 内部质量控制。

标准物质的回收率是准确度的替代度量。通过测量参比无烟烟草制品中的烟碱水平来确定回收率（表2）。回收率由下式计算：

$$回收率(\%) = 100 \times (分析结果/认证量)$$

表2 无烟烟草制品中烟碱含量的平均值和回收率

无烟烟草样品	认证值（mg/g）	平均烟碱含量（mg/g）	回收率（%）
CRP1	8.0	7.40	92.4
CRP2	12.0	10.58	88.2
CRP3	17.0	16.54	97.3
CRP4	9.0	9.03	100.4

21.3 分析特异性。

目标分析物的保留时间用于验证分析特异性。使用质量控制无烟烟草制品的内标物响应比范围来验证未知样品结果的特异性。

21.4 线性

建立的烟碱标准曲线在 50~1500 mg/L（1.3~40 mg/g）的标准浓度范围内呈线性。

21.5 潜在干扰。

丁香酚或调味物质的存在会引起干扰，因为其保留时间与烟碱相似。含有丁香或添加调味物质的样品最容易产生干扰。实验室可通过调整色谱仪参数来排除干扰。

22 重复性和再现性

2020年9月至2021年3月期间进行的一项国际合作研究涉及13个实验室和4个CRP无烟烟草制品，根据WHO TobLabNet方法验证方案和本SOP，得出了该方法的以下值。

根据 ISO 5725-1[**2.4**]和 ISO 5725-2[**2.5**]对测试结果进行统计分析，得到表3所示的数据。

表3 无烟烟草制品烟碱含量（mg/g）测定的精密度限值

参比烟草制品	n	$\hat{m}$	重复性限（r）	再现性限（R）
CRP1	9	7.40	0.49	2.92
CRP2	8	10.58	0.38	2.75
CRP3	7	16.54	0.50	2.77
CRP4	10	9.03	0.49	1.90

附录1 无烟烟草制品中烟碱含量测定的典型色谱图

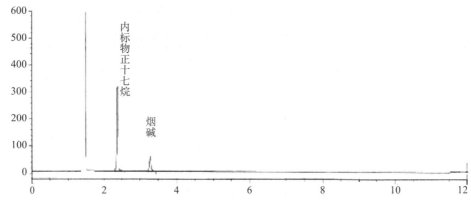

附图1 烟碱浓度为250 mg/L 的标准溶液的色谱图示例

WHO TobLabNet SOP 13

无烟烟草制品中水分含量的测定标准操作规程

方　　法：无烟烟草制品中水分含量的测定
基　　质：无烟烟草制品
更新时间：2021 年 12 月

无烟烟草制品正逐渐引起公共卫生组织的关注。2012 年在首尔举行的世界卫生组织《烟草控制框架公约》（WHO FCTC）缔约方大会第五次会议（COP5）要求确定管制无烟烟草制品中的化学成分的备选办法。本文件是应 2016 年德里举行的缔约方大会第七次会议（COP7）向 WHO FCTC 秘书处提出的邀请世界卫生组织根据 FCTC/COP6(12) 2b.ii 号决议中最终确定测量烟碱和烟草特定亚硝胺的标准操作规程的要求编写的。根据这一要求，WHO 组织了一项涉及其烟草实验室网络（TobLabNet）测试实验室的合作研究，测试了具有某些化学特性的材料，这些材料代表了一系列常见形式的无烟烟草制品，并且具有不同的物理和化学特性。本标准操作规程对经验证的 WHO 标准操作规程对无烟烟草制品的适用性进行了评价，并提出了推荐的方法。

本方法由世界卫生组织（WHO）烟草实验室网络（TobLabNet）成员实验室编写，作为无烟烟草制品中水分含量测定的标准操作规程（SOP）。水分含量是影响产品烟碱释放量的关键因素之一。

引言

为了建立具有可比性的无烟烟草制品水分含量的测定方法，并为 WHO TobLabNet 在全球范围内制定规程，需要无烟烟草制品特定参数的一致性检测方法。WHO TobLabNet 审查了常用操作来编制此 SOP，参考了 CORESTA 推荐方法 No. 76[2.1]和 AOAC 官方方法 966.02[2.2]。

1 适用范围

本方法规定了一种测定无烟烟草制品水分含量的烘箱干燥法。水分含量（烘箱挥发物）是在强制通风烘箱中，在 100℃ ± 1℃的温度条件下干燥样品 3 h ± 0.5 min 时质量的减少量。该方法可以测量无烟烟草制品的挥发性成分，包括在特定条件下损失的水和调味物质成分。准确测定无烟烟草制品中的水分含量至关重要，因为水分含量会影响产品的稳定性和完整性。

2 参考标准

2.1 CORESTA 推荐方法 No. 76：烟草和烟草制品水分含量（烘箱挥发物）的测定。

2.2 AOAC 官方方法 966.02：烟草水分重量测定法。

2.3 联合国毒品和犯罪问题办事处（UNODC）《代表性药品抽样指南》（维也纳，实验室和科学科，2009 年，http://www.unodc.org/documents/scientific/Drug_

Sampling.pdf)。

2.4 WHO TobLabNet SOP 02 烟草制品成分和释放物分析方法验证标准操作规程（世界卫生组织烟草实验室网络，日内瓦，2017 年）。

2.5 ISO 5725-1：检测方法和结果的准确度（正确度和精密度） 第 1 部分 一般原则和定义。

2.6 ISO 5725-2：检测方法和结果的准确度（正确度和精密度） 第 2 部分 确定标准检测方法重复性和再现性的基本方法。

3 术语和定义

3.1 水分含量：无烟烟草制品中的水分含量，以 mg/g 表示。

3.2 无烟烟草：无烟烟草制品中含有烟草的部分。

3.3 无烟烟草制品：完全或部分以烟叶为原料制成的产品，用于吸吮、咀嚼或鼻吸（世界卫生组织《烟草控制框架公约》第 1(f) 条），包括鼻烟（干鼻烟和湿鼻烟）、咀嚼烟草或源自烟草植物的混合材料。

3.4 实验室样品：用于实验室检测的样品，由一次性或在规定时间内送到实验室的单一类型产品组成。

3.5 试样：从实验室样品中随机抽取的待测样品。所取样品数量应能代表实验室样品。

3.6 试料：从试样中随机抽取的用于单次测试的样品。所取样品数量应能代表试样。

3.7 湿无烟烟草质量：无烟烟草制品放入烘箱之前的质量，以测试其水分含量。

3.8 干无烟烟草质量：无烟烟草制品放入烘箱干燥后的质量，以测试其水分含量。

4 方法概述

4.1 在本方法中，水分含量定义为样品置于通风烘箱中在 100℃ ± 1℃下干燥 3 h ± 0.5 min 时质量的减少量。

4.2 干燥后，将样品置于干燥器中冷却至室温，以防止从空气中吸收湿气。

5 安全和环境预防措施

5.1 遵守所有化学实验室活动的常规安全与环境预防措施。
5.2 使用本检测方法对某些产品进行检测和评估时所用材料或设备可能对

环境有害。本检测方法无力解决所有与使用相关的安全问题。所有使用此方法的人都有责任咨询相关机构并遵循现有监管要求，制定健康与安全规程以及环境预防措施。

5.3 应特别注意避免吸入或皮肤接触危险化学品。在制备或处理未稀释材料、标准溶液、稀释液或收集样品时，应在通风橱中进行，并穿戴合适的实验服、手套和护目镜。

6 仪器和设备

常规实验仪器，特别是：

6.1 可将空气温度保持在100℃±1℃的通风烘箱。

6.2 干燥器。

6.3 分析天平。

6.4 盛放样品的坩埚或蒸发皿，或者替代物。

7 试剂和耗材

超纯水。

8 玻璃器皿的准备

清洁、干燥的玻璃器皿，避免残留物污染。

9 溶液配制

不适用。

10 标准溶液配制

不适用。

11 抽样

11.1 按照实验室抽样程序对无烟烟草制品进行抽样。可根据各实验室惯例或者要求的特殊规定或样品可用性，使用替代方法来获取具有代表性的实验室样品。

11.2 试样组成

    11.2.1 若可能，将实验室样品分成独立的单元（如包、条）。

    11.2.2 从至少 $\sqrt{n}$ 个单元[2.3]中取出每种试样的等量产品。

12 样品准备

12.1 将无烟烟草制品从包或条中取出。包括质量控制样本（如适用）。

12.2 根据个别实验室惯例（例如可采用食品分析抽样方法），从无烟烟草制品中选取适量且具有代表性的部分。

12.3 从无烟烟草制品中提取无烟烟草。

12.4 将足够的无烟烟草样品混合，使每个试料约含 0.5~2 g 均质无烟烟草。

13 吸烟机设置

不适用。

14 样品生成

不适用。

15 样品制备

15.1 烘箱准备。

打开烘箱，并将温度设置为 100℃ ± 1℃。使用前让烘箱平衡至少 1 h，并确保温度稳定在 100℃ ± 1℃。

15.2 样品制备。

    15.2.1 在分析天平上称量干净、干燥的蒸发皿，并将质量记录为 W_T。

    15.2.2 从天平上取下蒸发皿，根据尺寸，加入 1~5 g 无烟烟草试样。

    15.2.3 将装有样品的蒸发皿置于天平上，记录质量 W_1，精确至 0.01 g。

    15.2.4 检查烘箱温度是否为 100℃ ± 1℃，并将样品放入烘箱中。

    15.2.5 关闭烘箱门，记录开始时间，干燥 3 h ± 0.5 min。

    15.2.6 对所有样品重复步骤 **15.2.1~15.2.5**。

16 样品分析

16.1 干燥 3 h ± 0.5 min 后，将装有样品的蒸发皿从烘箱中取出，放入干燥器中。

16.2 使样品在干燥器中冷却约 30 min 至室温。

16.3 在天平上称量蒸发皿中的样品。质量记录为 W_2，精确至 0.01 g。

16.4 对每个样品重复步骤 **16.3**。

17 数据分析与计算

17.1 使用下式计算含水量百分比：

$$M(\%) = \frac{W_1 - W_2}{W_1 - W_T} \times 100$$

式中：M——含水率；
W_1——无烟烟草样品和蒸发皿的初始质量；
W_2——干燥后的无烟烟草样品和蒸发皿的质量；
W_T——蒸发皿的皮重。

计算的水分含量可使用下式将原样或湿重基础上的分析物浓度转换为干重基础上的分析物浓度：

$$c_{dry} = c_{wet} \times \frac{100}{100 - M}$$

式中：M——含水率（%）；
c_{dry}——干重基础上的分析物浓度；
c_{wet}——原样或湿重基础上的分析物浓度。

18 特别注意事项

不适用。

19 数据报告

19.1 报告每个评估样品的单独测量值。

19.2 按照方法规范的规定报告结果。

19.3 更多信息可参见 WHO TobLabNet SOP 02[2.4]。

20 质量控制

20.1 控制参数。

注 1：如果控制测量值超出预期值的公差极限，则必须进行适当的调查并采取措施。

注 2：如有必要，需要做额外的实验室质量保证规程，以符合各个实验室的做法。

20.2 质量控制样品。

为了验证整个分析的一致性，根据各实验室的实践分析参比烟草制品，如 CORESTA 参比产品（CRP）。

21 方法性能

21.1 报告限。

报告限设置为最低含水量，以质量百分比的形式确定。

21.2 内部质量控制。

标准物质的回收率是准确度的替代度量。通过测量参比无烟烟草制品中的水分含量来确定回收率（表 1）。回收率由下式计算：

$$回收率(\%) = 100 \times (分析结果/认证量)$$

表 1 无烟烟草制品中水分含量的平均值和回收率

无烟烟草样品	认证值（%）	平均含水量（%）	回收率（%）
CRP1	53.7	53.87	100.3
CRP2	53.0	51.20	96.6
CRP3	8.1	7.67	94.7
CRP4	23.0	24.07	104.7

22 重复性和再现性

2020 年 9 月至 2021 年 3 月期间进行的一项国际合作研究涉及 13 个实验室和 4 个 CRP 无烟烟草制品，根据 WHO TobLabNet 方法验证方案和本 SOP，得出了该方法的以下值。

根据 ISO 5725-1[2.5]和 ISO 5725-2[2.6]对测试结果进行统计分析，得到表 2 所示的数据。

表 2　无烟烟草制品中水分含量（%）测定的精密度限值

参比烟草制品	n	平均值	重复性限（r）	再现性限（R）
CRP1.1	9	53.87	1.02	1.93
CRP2.1	10	51.20	0.76	2.73
CRP3.1	10	7.67	0.81	2.68
CRP4.1	12	24.07	1.82	4.78

WHO TobLabNet SOP 14

无烟烟草制品 pH 的测定标准操作规程

方　　法：无烟烟草制品 pH 的测定
分　析　物：pH
基　　质：无烟烟草制品
更新时间：2021 年 12 月

无烟烟草制品正逐渐引起公共卫生组织的关注。2012年在首尔举行的世界卫生组织《烟草控制框架公约》（WHO FCTC）缔约方大会第五次会议（COP5）要求确定管制无烟烟草制品中的化学成分的备选办法。本文件是应2016年德里举行的缔约方大会第七次会议（COP7）向 WHO FCTC 秘书处提出的邀请世界卫生组织根据 FCTC/COP6(12) 2b.ii 号决议中最终确定测量烟碱和烟草特定亚硝胺的标准操作规程的要求编写的。根据这一要求，WHO 组织了一项涉及其烟草实验室网络（TobLabNet）测试实验室的合作研究，测试了具有某些化学特性的材料，这些材料代表了一系列常见形式的无烟烟草制品，并且具有不同的物理和化学特性。本标准操作规程对经验证的 WHO 标准操作规程对无烟烟草制品的适用性进行了评价，并提出了推荐的方法。

本方法由世界卫生组织（WHO）烟草实验室网络（TobLabNet）成员实验室编写，作为无烟烟草制品 pH 测定的标准操作规程（SOP）。pH 是影响产品烟碱释放量的关键因素之一。

引言

为了在全球范围内建立具有可比性的烟草制品测定方法，需要特定参数的一致性检测方法。WHO TobLabNet 审查了测定无烟烟草制品 pH 的常用操作来编制此 SOP，参考了 CORESTA 推荐方法 No. 69 [2.1]，用于描述无烟烟草制品 pH 测定的方法。

1 适用范围

本方法提供了通过水提取和 pH 计读数测定无烟烟草制品 pH 的一般指南。该方法适用于 pH 在 4~14 范围内的无烟烟草制品 pH 的测定。

2 参考标准

2.1 CORESTA 推荐方法 No. 69：烟草和烟草制品 pH 的测定。

2.2 ISO 3696:1987 分析实验室用水 规范和试验方法。

2.3 联合国毒品和犯罪问题办公室（UNODC）《代表性药品抽样指南》（维也纳，实验室和科学科，2009年，http://www.unodc.org/documents/scientific/Drug_Sampling.pdf）。

2.4 WHO TobLabNet SOP 02 烟草制品成分和释放物分析方法验证标准操作规程（世界卫生组织烟草实验室网络，日内瓦，2017年）。

2.5 ISO 5725-1：检测方法和结果的准确度（正确度和精密度） 第1部分

一般原则和定义。

2.6 ISO 5725-2：检测方法和结果的准确度（正确度和精密度） 第2部分 确定标准检测方法重复性和再现性的基本方法。

3 术语和定义

3.1 pH：无烟烟草制品的pH。

3.2 无烟烟草：无烟烟草制品中含有烟草的部分。

3.3 无烟烟草制品：完全或部分以烟叶为原料制成的产品，用于吸吮、咀嚼或鼻吸（世界卫生组织《烟草控制框架公约》第1(f)条），包括鼻烟（干鼻烟和湿鼻烟）、咀嚼烟草或源自烟草植物的混合材料。

3.4 实验室样品：用于实验室检测的样品，由一次性或在规定时间内送到实验室的单一类型产品组成。

3.5 试样：从实验室样品中随机抽取的待测样品。所取样品数量应能代表实验室样品。

3.6 试料：从试样中随机抽取的用于单次测试的样品。所取样品数量应能代表试样。

4 方法概述

制备无烟烟草制品样品的水萃取物，并用pH电极测量其pH。

5 安全和环境预防措施

5.1 遵守所有化学实验室活动的常规安全与环境预防措施。

5.2 使用本检测方法对某些产品进行检测和评估时所用材料或设备可能对环境有害。本检测方法无力解决所有与使用相关的安全问题。所有使用此方法的人都有责任咨询相关机构并遵循现有监管要求，制定健康与安全规程以及环境预防措施。

5.3 应特别注意避免吸入或皮肤接触危险化学品。在制备或处理未稀释材料、标准溶液、稀释液或收集样品时，应在通风橱中进行，并穿戴合适的实验服、手套和护目镜。

6 仪器和设备

常规实验仪器，特别是：

6.1 pH 计。

6.2 使萃取瓶保持在适当位置的线性振荡器。

6.3 萃取瓶：带有塞子的锥形瓶（100 mL），带有波纹密封盖和隔层的隔热玻璃瓶（100 mL），带有聚四氟乙烯塞子的培养管或其他合适的烧瓶。

7 试剂和耗材

除非另有说明，所有试剂应至少为分析纯。试剂尽可能以 CAS 号标识。

7.1 水：符合 ISO 3696:1987[**2.2**] 2 级或更高标准。

7.2 标准 pH 缓冲液（pH＝4.01, 9.21 或 ＞10，根据实验室 pH 标准溶液的可用性而定）。

8 玻璃器皿的准备

清洁、干燥的玻璃器皿，避免残留物污染。

9 溶液配制

不适用。

10 标准溶液配制

不适用。

11 抽样

11.1 按照实验室抽样程序对无烟烟草制品进行抽样。可根据各实验室惯例或者要求的特殊规定或样品可用性，使用替代方法来获取具有代表性的实验室样品。

11.2 试样组成

11.2.1 若可能，将实验室样品分成独立的单元（如包、条）。

11.2.2 从至少 $\sqrt{n}$ 个单元[**2.3**]中取出每种试样的等量产品。

12 样品准备

12.1 将无烟烟草制品从包或条中取出。包括质量控制样本（如适用）。

12.2 根据个别实验室惯例（例如可采用食品分析抽样方法），从无烟烟草制品中选取适量且具有代表性的部分。

12.3 从无烟烟草制品中提取无烟烟草。

12.4 将足够的无烟烟草样品混合，使每个试料约含 2~5 g 均质无烟烟草。

13 吸烟机设置

不适用。

14 样品生成

不适用。

15 样品制备

15.1 称取至少 1.0 g ± 0.1 g 无烟烟草样品，置于 100 mL 锥形瓶中。建议该方法的样品质量范围为 1~2 g。

15.2 向锥形瓶中加入 20.0 mL ± 0.5 mL 水，轻轻摇动约 30 min。

16 样品分析

16.1 pH 计的校准。

  16.1.1 用至少两种 pH 缓冲溶液（pH = 4.01 和 9.21 或以上）校准 pH 电极；在校准范围内进行两点校准。校准和样品的测量需要在相同的温度下连续进行。

  16.1.2 在电极可用于样品测量之前，电极斜率应在计算值的 95%~105% 范围内。

  16.1.3 每次测量前后，用蒸馏水冲洗电极。

16.2 样品 pH 测定。

  16.2.1 在停止摇动后 60 min 内测量无烟烟草样品的 pH。

  16.2.2 摇匀后，让样品瓶静置 20 min 使溶液沉淀，用 Whatman 41 号

纸过滤，或在必要时以 10000 r/min 中速离心。

- **16.2.3** 记录温度。所有 pH 测量应在室温 20~25℃±1℃下进行。
- **16.2.4** pH 测量值精确至小数点后两位。
- **16.2.5** 样品的 pH 必须每隔 15 min 重复测量一次。报告 1 h（4 个读数）的平均测量值。读数之间允许的最大 pH 变化为 0.3 个单位。
- **16.2.6** 每次测量前后用蒸馏水冲洗电极。

17 数据分析与计算

不适用。

18 特别注意事项

不适用。

19 数据报告

19.1 测试报告应提供小数点后两位的 pH 测量结果。
19.2 更多信息可参见 WHO TobLabNet SOP 02[2.4]。

20 质量控制

20.1 控制参数。
注 1：如果控制测量值超出预期值的公差极限，则必须进行适当的调查并采取措施。
注 2：如有必要，需要做额外的实验室质量保证规程，以符合各个实验室的做法。
20.2 质量控制样品。
为了验证整个分析的一致性，根据各实验室的实践分析参比烟草制品，如 CORESTA 参比产品（CRP）。

21 方法性能

标准物质的回收率是准确度的替代度量。通过测量参比无烟烟草制品中的 pH

来确定回收率（表1）。回收率由下式计算。

回收率(%) = 100×(分析结果/认证量)

表1 无烟烟草制品中 pH 的平均值和回收率

无烟烟草样品	认证值	平均 pH	回收率（%）
CRP1	8.4	7.68	92.8
CRP2	7.7	8.05	103.6
CRP3	7.1	7.04	99.5
CRP4	6.2	5.95	95.7

22 重复性和再现性

2020 年 9 月至 2021 年 3 月期间进行的一项国际合作研究涉及 13 个实验室和 4 个 CRP 无烟烟草制品，根据 WHO TobLabNet 方法验证方案和本 SOP，得出了该方法的以下值。

根据 ISO 5725-1[2.5]和 ISO 5725-2[2.6]对测试结果进行统计分析，得到表 2 所示的数据。

表2 无烟烟草制品 pH 测定的精密度限值

参比烟草制品	n	平均值	重复性限（r）	再现性限（R）
CRP1.1	13	7.68	0.17	1.70
CRP2.1	13	8.05	0.20	1.31
CRP3.1	13	7.04	0.10	2.84
CRP4.1	13	5.95	0.10	1.00

WHO TobLabNet SOP 15

加热型烟草制品中烟碱、甘油和丙二醇含量的测定标准操作规程

方　　法：	加热型烟草制品中烟碱、甘油、丙二醇含量的测定
分 析 物：	烟碱（3-[(2S)-1-甲基吡咯烷基-2-基]吡啶）（CAS 号：54-11-5）
	甘油（丙烷-1,2,3-三醇）（CAS 号：56-81-5）
	丙二醇（丙烷-1,2-二醇）（CAS 号：57-55-6）
	三乙酸甘油酯（CAS 号：102-76-1）
基　　质：	烟草
更新时间：	2023 年 8 月 25 日

本方法由世界卫生组织（WHO）健康促进司无烟草行动组和 WHO 烟草实验室网络（TobLabNet）成员编制，作为加热型烟草制品（HTP）中烟碱、甘油和丙二醇含量的测定标准操作规程（SOP）。该方法也适用于在实验室进行适当验证后的三乙酸甘油酯的定量，特别注意推荐的质量控制标准。

引言

为了在全球范围内建立具有可比性的加热型烟草制品检测方法，需要加热型烟草制品中烟碱、甘油和丙二醇含量测定的一致性检测方法。WHO TobLabNet 审查了测定加热型烟草制品中烟碱、甘油和丙二醇含量的常规操作，以编制本标准操作规程（SOP）。

2018 年 10 月 1~6 日在瑞士日内瓦举行的世界卫生组织《烟草控制框架公约》（WHO FCTC）缔约方大会第八次会议（COP8）要求公约秘书处邀请 WHO 和 WHO TobLabNet 遵照关于新型烟草制品的 FCTC/COP8(22)号决议第 2 段所述：①评估现有的成分和释放物标准操作规程是否适用或经过调整后适用于加热型烟草制品；②酌情就测量这些产品的成分和释放物的合适方法提出建议。

本标准操作规程根据 WHO TobLabNet SOP 11 [2.1]，WHO TobLabNet SOP 06 [2.2]和 Chen 等的出版物[2.3]进行了修改，描述了加热型烟草制品中烟碱、甘油和丙二醇含量的测定方法。

1 适用范围

本方法适用于气相色谱法（GC）测定加热型烟草制品中烟碱、甘油和丙二醇的含量，也可用于测定三乙酸甘油酯，但由于报道的结果数量有限，尚未通过实验验证。本方法的工作范围对烟碱可达 50 mg/g，对甘油可达 500 mg/g，对丙二醇高达 100 mg/g，对三乙酸甘油酯高达 100 mg/g（可选）。

2 参考标准

2.1 WHO TobLabNet SOP 11 电子烟烟液中烟碱、甘油和丙二醇含量的测定标准操作规程（世界卫生组织烟草实验室网络，日内瓦，2021 年）。

2.2 WHO TobLabNet SOP 06 卷烟烟丝中保润剂的测定标准操作规程（世界卫生组织烟草实验室网络，日内瓦，2016 年）。

2.3 Chen A X, Akmam Morsed F and Cheah N P. 2021. A Simple Method to Simultaneously Determine the Level of Nicotine, Glycerol, Propylene Glycol, and Triacetin in Heated Tobacco Products by Gas Chromatography Flame Ionization

Detection. Journal of AOAC INTERNATIONAL.

2.4 ISO 13276：烟草和烟草制品 烟碱纯度的测定 钨硅酸重量法。

3 术语和定义

3.1 烟碱、甘油、丙二醇和三乙酸甘油酯（可选）含量：加热型烟草制品中烟碱、甘油、丙二醇（和三乙酸甘油酯）的各自含量，以 mg/g HTP 烟草表示。

3.2 加热型烟草制品（HTP）：一种含有烟草或烟草基质，由单一热源（如电、气溶胶、碳等）加热，以产生含烟碱的气溶胶供使用者吸入的设备。

3.3 实验室样品：用于进行实验室检测的样品，由一次性或在规定时间内送到实验室的单一类型产品组成。

3.4 试样：从实验室样品中随机抽取的待测样品。所取样品数量应能代表实验室样品。

3.5 试料：从试样中随机抽取的用于单次测试的样品。所取样品数量应能代表试样（用于分析的试样的数量或体积，通常为 IUPAC 定义的已知质量或体积）。

注：每个试样分析的试料数量应适应样品的不均匀性。

4 方法概述

4.1 从 HTP 烟草填料中提取烟碱、甘油和丙二醇，以 70%甲醇/30%乙腈[7.7][7.8]和内标物[7.9][7.10]为萃取剂。采用气相色谱-氢火焰离子化检测器（GC-FID）测定分析物含量。

4.2 通过将分析物（烟碱、甘油、丙二醇）与相应内标物的色谱图峰面积比进行作图，建立已知浓度分析物的校准曲线，从而确定试料的分析物含量。

5 安全和环境预防措施

5.1 遵守所有化学实验活动的常规安全与环境预防措施。

5.2 使用本检测方法对某些产品进行检测和评估时所用材料或设备可能对环境有害。本检测方法无力解决所有与使用相关的安全问题。所有使用此方法的人都有责任咨询相关机构并遵循现有监管要求，制定健康与安全规程以及环境预防措施。

5.3 应特别注意避免吸入或皮肤接触危险化学品。在制备或处理未稀释材料、标准溶液、稀释液或收集样品时，应在通风橱中进行，并穿戴合适的实验服、手套和护目镜。

6 仪器和设备

常规实验仪器，特别是：

6.1 配置用于保持容器位置的声呐器。

6.2 配备氢火焰离子化检测器的气相色谱仪（GC-FID）。

6.3 能够分离溶剂、内标物、烟碱和其他成分色谱峰的毛细管气相色谱柱（如 Agilent DB-ALC1，30 m × 0.32 mm, 1.8 μm）。

6.4 经校准的分析天平，精密度为 0.0001 g。

6.5 吸耳球和注射器吸液管（1.0 mL 至 20.0 mL，适用于样品和标准品制备）。

6.6 A 级容量瓶（适用于标准品制备）。

6.7 标准的食品研磨机。

7 试剂和耗材

除非另有说明，所有试剂应至少为分析纯。试剂尽可能以 CAS 号识别。

7.1 载气：高纯度（＞99.999%）的氦气（CAS 号：7440-59-7）。高纯度（＞99.999%）的氢气（CAS 号：1333-74-0）可作为替代载气。

7.2 辅助气体：用于氢火焰离子化检测器的高纯度（＞99.999%）的空气和氢气（CAS 号：1333-74-0）。

7.3 纯度不低于98%的烟碱（CAS 号：54-11-5）[2.4]。也可使用纯度不低于98%的水杨酸烟碱盐盐（CAS 号：29790-52-1）。

7.4 纯度不低于98%的甘油（CAS 号：56-81-5）。

7.5 纯度不低于98%的丙二醇（CAS 号：57-55-6）。

7.6 纯度不低于98%的三乙酸甘油酯（CAS 号：102-76-1）（可选）。

7.7 甲醇，色谱纯度（CAS 号：67-56-1）。

7.8 乙腈，色谱纯度（CAS 号：75-05-8）。

7.9 烟碱和三乙酸甘油酯的内标物（可选）：纯度不低于 98%的正十七烷（CAS 号：629-78-7）。

7.10 甘油和丙二醇的内标物：纯度不低于99%的1,3-丁二醇（CAS 号：107-88-0）。

8 玻璃器皿的准备

清洁、干燥的玻璃器皿，避免残留物污染。

9 溶液配制

9.1 萃取液

萃取液由 70%甲醇/30%乙腈组成[7.7][7.8]（700 mL 甲醇加 300 mL 乙腈），含适量内标物：

将 0.50 mL 正十七烷[7.9]加 2.00 mL 1,3-丁二醇[7.10]移液至 1 L 容量瓶中。

用 70%甲醇/30%乙腈[7.7][7.8]定容至 1 L，充分混合后转移至装有防止污染功能的储存容器中。

注：考虑内标物对灵敏度和选择性的影响以及方法的线性范围，内标物的浓度和/或类型可以调整。

10 标准溶液配制

下述标准溶液配制方法仅供参考，如有需要可进行调整。在低温下储存的溶剂和溶液在使用前应平衡至(22 ± 5)℃。

10.1 烟碱标准储备液（5 g/L）

称取约 500 mg 烟碱[7.3]或 925 mg 水杨酸烟碱盐（精确至 0.1 mg），置于 100 mL 容量瓶中，用萃取液[9.1]定容至刻度。

混合均匀，并在 0~4℃条件下避光储存。

10.2 甘油标准储备液（50 g/L）

称取约 5000 mg 甘油[7.4]，精确至 0.1 mg，置于 100 mL 容量瓶中，用萃取液[9.1]定容至刻度。

混合均匀，并在 0~4℃条件下避光储存。

10.3 丙二醇标准储备液（5 g/L）

称取约 500 mg 丙二醇[7.5]，精确至 0.1 mg，置于 100 mL 容量瓶中，用萃取液[9.1]定容至刻度。

混合均匀，并在 0~4℃条件下避光储存。

10.4 三乙酸甘油酯储备液（20 g/L）（可选）

称约 2000 mg 三乙酸甘油酯[7.6]，精确至 0.1 mg，置于 100 mL 容量瓶中，用萃取液[9.1]定容至刻度。

混合均匀，并在 0~4℃条件下避光储存。

10.5 标准溶液

 10.5.1 用移液管将 **10.1** 中配制的指定量烟碱储备液移液至 100 mL 容量瓶中，如表 1 所示。

 10.5.2 用移液管将 **10.2** 中配制的指定量甘油储备液移液至相同的 100

mL 容量瓶中[**10.5.1**]，如表 2 所示。

10.5.3 用移液管将 **10.3** 中配制的指定量丙二醇储备液移液至相同的 100 mL 容量瓶中[**10.5.1**]，如表 3 所示。

（可选）将 **10.4** 配制的三乙酸甘油酯储备液移液至相同的 100 mL 容量瓶中[**10.5.1**]，如表 4 所示。

10.5.4 向容量瓶中加入萃取液[**9.1**]，定容至刻度（100 mL）。

10.5.5 标准溶液避光保存在 4~8℃。

10.5.6 标准溶液的烟碱、甘油、丙二醇和三乙酸甘油酯（可选）最终浓度由下式确定：

$$最终浓度（mg/mL）= \frac{x \cdot y}{10000}$$

式中：x——称取的分析物[**10.1**]、[**10.2**]、[**10.3**]或[**10.4**]的原始质量（mg）；

y——移取的储备液[**10.5.1**]、[**10.5.2**]、[**10.5.3**]和[**10.5.4**]的体积（mL）。

标准溶液中烟碱的浓度如表 1 所示，甘油浓度如表 2 所示，丙二醇浓度如表 3 所示，三乙酸甘油酯浓度如表 4 所示（可选）。

表 1 标准溶液中烟碱的浓度

标准溶液	烟碱储备液[10.1] 体积（mL）	总体积（mL）	标准溶液中烟碱最终浓度 （mg/mL）
1	2.0	100	0.1
2	4.0	100	0.2
3	8.0	100	0.4
4	16.0	100	0.8
5	20.0	100	1.0

表 2 标准溶液中甘油的浓度

标准溶液	甘油储备液[10.2] 体积（mL）	总体积（mL）	标准溶液中甘油最终浓度 （mg/mL）
1	1.0	100	0.5
2	4.0	100	2.0
3	8.0	100	4.0
4	16.0	100	8.0
5	20.0	100	10.0

表3 标准溶液中丙二醇的浓度

标准溶液	丙二醇储备液[10.3]体积（mL）	总体积（mL）	标准溶液中丙二醇最终浓度（mg/mL）
1	0.6	100	0.0
2	4.0	100	0.2
3	10.0	100	0.5
4	20.0	100	1.0
5	40.0	100	2.0

表4 标准溶液中三乙酸甘油酯的浓度（可选）

标准溶液	三乙酸甘油酯储备液[10.4]体积（mL）	总体积（mL）	标准溶液中三乙酸甘油酯最终浓度（mg/mL）
1	0.5	100	0.1
2	1.5	100	0.3
3	2.5	100	0.5
4	5.0	100	1.0
5	10.0	100	2.0

标准溶液范围可根据所使用的设备和待测样品进行调整，同时要考虑方法灵敏度可能对结果产生的影响。

在使用前，应将所有溶剂和溶液调节至室温(22 ± 5)℃。

11 抽样

11.1 样品收集。

根据适用的法规，并考虑样品的可用性，选取具有代表性的HTP样品。

11.2 试样组成。

若可能，将实验室样品分成独立的单元（如包、条）。

对于每个试样，从至少$\sqrt{n}$个单元[2.5]中取出具有代表性数量的等量产品。

12 样品准备

12.1 从HTP烟棒的至少一个销售单元中取出烟草，从中形成试料和质量控制样品（如适用）。

12.2 将从12.1取出的HTP烟棒中的烟草混合至不少于2 g并均质。

13 吸烟机设置

不适用。

14 样品生成

不适用。

15 样品制备

15.1 将 **12.2** 中的烟草混合研磨直至均匀。
15.2 将 0.2~0.3 g 混合均匀的烟草样品放入玻璃提取容器中。
15.3 向样品中加入 10 mL 萃取液。
15.4 对容器进行 60 min 的超声处理。
15.5 将等量的样品萃取物转移到自动进样器样品瓶中,同时防止固体颗粒进入。
15.6 如果样品在分析前要储存,应在 4~8℃避光条件下储存。

16 样品分析

使用配备氢火焰离子化检测器的气相色谱仪测定加热型烟草制品中的烟碱、甘油、丙二醇和三乙酸甘油酯(可选)。分析物可在所用的色谱柱上与其他潜在干扰物分离开。通过比较测试溶液的峰面积比与已知浓度标准溶液的峰面积比,得到分析物的浓度。

16.1 GC 操作条件示例。

GC 柱:Agilent DB-ALC1(30 m × 0.32 mm,1.8 μm)或同等产品。
烘箱:140℃,5 min;
　　　140~180℃,40℃/min;
　　　80℃,4 min;
　　　180~230℃,5℃/min;
进样口温度:225℃;
检测器温度:260℃;
载气:流速为 1.5 mL/min 的氦气;
进样体积:1.0 μL;
进样方式:分流。
注:根据仪器和色谱柱条件以及色谱峰的分离度调整操作参数。

在上述条件下,预计总分析时间约为 20 min(延长分析时间可以优化性能)。如果使用气相色谱-质谱(GC-MS)作为 GC-FID 的替代方法,附录 2 提供了 GC-MS 设置的示例。

16.2 预计保留时间。

在分析开始之前，必须验证洗脱顺序和保留时间。

注：对于 **16.1** 中描述的条件，预期的洗脱顺序是丙二醇、1,3-丁二醇、甘油、烟碱、三乙酸甘油酯和正十七烷。

温度、气体流速和色谱柱的使用年限的差异可能会改变保留时间。

16.3 烟碱、甘油、丙二醇、三乙酸甘油酯的测定（可选）。

测试样品的测量顺序应按照适用的实验室质量体系设计。本节给出一个确定 HTP 烟草中烟碱、甘油、丙二醇和三乙酸甘油酯（可选）含量的方法和测定顺序的示例。

在相同的条件下注入等量的标准溶液和测试溶液。

16.3.1 在使用前通过注入两份 1 μL 的样品溶液作为预试样来调节系统。

16.3.2 在与样品相同的条件下注入 1 μL 萃取液[9.1]和一份标准溶液，验证气相色谱系统的性能和所用试剂的成分污染情况。

16.3.3 将烟碱、甘油、丙二醇和三乙酸甘油酯（可选）的标准溶液注入气相色谱仪。

16.3.4 评估标准溶液的保留时间和响应（面积计数）。如果保留时间与之前进样的保留时间相似（±0.2 min），并且响应值与之前进样的典型响应值的差异在 20% 以内，则系统已准备就绪，可进行分析。如果保留时间和/或响应值超出规范范围，则应根据各实验室政策采取纠正措施。

16.3.5 记录烟碱、甘油、丙二醇和三乙酸甘油酯（可选）及内标物的峰面积。

16.3.6 根据下式计算标准溶液（包括空白溶剂）中每种分析物的分析物峰与内标峰的峰面积比。

$$RF = A_{analyte} / A_{IS}$$

式中：RF——峰面积比；

$A_{analyte}$——分析物峰的峰面积；

A_{IS}——内标物峰的峰面积。

16.3.7 绘制标准溶液中分析物的浓度（X 轴）和 **16.3.6** 中计算的峰面积比（RF）（Y 轴）的曲线图。

注：标准曲线应在整个范围内应呈线性。

16.3.8 使用线性回归的斜率（b）和截距（a）来计算线性回归方程（$Y = a + bx$）。如果 $R^2 < 0.99$，则应重新校准。根据实验室程序检查个别异常值。

16.3.9 注入 1 μL 每种试料[15.5]和质量控制样品[20.3]，用相应软件测定峰面积。

注：代表性色谱图见附录1。

17 数据分析与计算

17.1 对于每个试料，计算分析物峰面积与内标峰面积的比值（Y_t）。

注：所有试料中的分析物峰面积比（Y_t）必须在校准曲线的工作范围内；否则，应根据需要调整标准溶液或试料浓度。

17.2 使用 **16.3.8** 中确定的校准曲线的系数，计算每个试料中分析物的浓度（mg/mL）。

$$M_t = (Y_t - a) / b$$

式中：M_t——试料中分析物的浓度（mg/mL）；
　　　Y_t——分析物峰与内标物峰面积比；
　　　a——**16.3.8** 线性回归得到的校准曲线截距；
　　　b——**16.3.8** 线性回归得到的校准曲线斜率。

17.3 使用下式计算以 mg/g 烟草表示的烟草试样中的分析物含量（m_c）:

$$m_c = \frac{m_t \times V_e}{m_o}$$

式中：m_c——试样中被分析物的含量（mg/g）；
　　　M_t——试料中分析物的浓度（mg/mL）；
　　　V_e——所用的萃取液的体积（mL）；
　　　m_o——试料的质量[12.1]（g）。

18 特别注意事项

18.1 安装新色谱柱后，在 **16.1** 中所述的 GC 条件下注入试样（烟草样品）溶液。应重复进样，直到该组分和内标物的峰面积（或峰高）可重现。这需要大约 4 次进样。

18.2 每组（序列）样品结束后，建议将色谱柱温度升高至 220℃ 30 min，从气相色谱柱中清除高沸点成分。

18.3 当内标物的峰面积（或峰高）明显高于预期时，建议用不加内标物的含有 70%甲醇/30%乙腈[7.7][7.8]的溶液提取等量的试料溶液，并按本操作规程进行分析。这样就可以确定是否有其他成分与内标物共洗脱，从而导致分析物数值偏低的误差。

19 数据报告

19.1 报告每个评估样品的单独测量值。
19.2 按照方法规范的要求报告结果。
19.3 更多信息可参见 WHO TobLabNet SOP 02[**2.6**]。

20 质量控制

20.1 控制参数。
注 1：如果控制测量值超出预期值的公差极限，则必须根据实验室质量程序进行适当的调查并采取措施。
注 2：如有必要，需要做额外的实验室质量保证规程，以符合各个实验室的做法。
20.2 实验室试剂空白样。
为了在样品制备和分析过程中检测潜在的污染，需包括萃取液的空白测定[**9.1**]。结果应小于相应分析物的检出限。
20.3 质量控制样品。
为了验证整个分析的一致性，根据各实验室的实践分析参比或质量控制 HTP 烟草样品。

21 方法性能

注：鼓励实验室按照其质量规范对本方法进行验证。
21.1 检测限（LOD）和定量限值（LOQ）。
LOD 设定为空白测定平均值标准差的 3 倍，LOQ 设定为空白测定平均值标准差的 10 倍。LOD 和 LOQ 如表 5 所示（数据来自单个实验室验证）。

表 5　加热型烟草制品中烟碱、甘油、丙二醇和三乙酸甘油酯的 LOD 和 LOQ

分析物	LOD（mg/g）	LOQ（mg/g）
烟碱	0.05	0.2
甘油	1.4	4.7
丙二醇	1.0	1.0
三乙酸甘油酯	0.2	0.8

注：表 5 中列出的 LOD 值和 LOQ 值超出了本方法的工作范围

21.2 实验室基质加标回收率。
加入到基质上的分析物的回收率用作准确度的替代量度。采用单一实验室验证法测定各成分的回收率。由于没有不含烟碱、甘油、丙二醇和三乙酸甘油酯的生烟,所以用茶叶进行回收率的测定。如果使用生烟草测定回收率,结果将超出本方法的工作范围。

将不同量的烟碱、甘油、丙二醇和三乙酸甘油酯储备液分别加入到含有 0.2~0.3 g 茶叶的 10 mL 容量瓶中,通过单一实验室验证测定回收率。超声均质化处理 60 min 后[**6.1**],将回收样品转移到自动进样器样品瓶中。对于每个加标样品,通过在一天内分析单个小瓶来测定烟碱、甘油、丙二醇和三乙酸甘油酯。第二天重复这个过程,得到如下结果。回收率由下式计算,烟碱、甘油、丙二醇、三乙酸甘油酯的回收率分别如表6至表9所示。

$$R = 100 \times \frac{c_r - c_n}{c}$$

式中:R——回收率(%);
c_r——回收样品中分析物的浓度(mg/mL);
c_n——天然样品中分析物的浓度(mg/mL);
c——回收样品中的标称分析物浓度(mg/mL)。

表6 通过单个实验室验证的烟碱的平均值和回收率

第一天			第二天		
加标量(mg/g)	平均值(mg/g)	回收率(%)	加标量(mg/g)	平均值(mg/g)	回收率(%)
5.2	5.0	89.8	5.3	5.2	91.6
20.6	20.2	97.0	21.1	20.8	95.9
41.3	40.8	97.4	42.2	41.0	95.0

表7 通过单个实验室验证的甘油的平均值和回收率

第一天			第二天		
加标量(mg/g)	平均值(mg/g)	回收率(%)	加标量(mg/g)	平均值(mg/g)	回收率(%)
25.1	25.0	98.1	25.2	25.4	100.5
251.0	250.7	98.2	251.4	244.1	95.2
401.6	409.2	101.1	402.3	398.2	97.2

表8 通过单个实验室验证的丙二醇的平均值和回收率

第一天			第二天		
加标量(mg/g)	平均值(mg/g)	回收率(%)	加标量(mg/g)	平均值(mg/g)	回收率(%)
2.5	2.5	98.8	2.6	2.5	98.6
25.3	25.1	98.0	25.5	24.8	95.5
75.8	76.1	100.2	76.6	77.4	100.9

表 9　通过单个实验室验证的三乙酸甘油酯的平均值和回收率

第一天			第二天		
加标量（mg/g）	平均值（mg/g）	回收率（%）	加标量（mg/g）	平均值（mg/g）	回收率（%）
5.0	4.9	90.6	5.0	5.0	98.8
50.3	49.6	93.3	50.2	50.3	98.4
150.9	153.8	99.7	150.7	151.8	100.4

21.3 分析特异性。

目标分析物的保留时间用于验证分析特异性。分析物与质量控制烟草内部标准响应比的范围用于验证未知样品结果的特异性。

21.4 线性。

建立的烟碱校准曲线在 0.1~1.0 mg/mL 标准浓度范围内呈线性。

甘油在 0.5~10.0 mg/mL 标准浓度范围内呈线性。

丙二醇在 0.03~2.0 mg/mL 标准浓度范围内呈线性。

三乙酸甘油酯在 0.1~2.0 mg/mL 标准浓度范围内呈线性（可选）。

21.5 潜在干扰。

由于与分析物或内标物的保留时间相似，调味物质的存在会引起干扰。

22　重复性和再现性

2022 年 1~6 月进行的一项国际合作研究涉及 12 个实验室和 3 个加热型烟草制品样品，给出了本方法的以下值。

根据 ISO 5725-1[2.7]和 ISO 5725-2[2.8]对试验结果进行了统计分析，从而给出了表 10 至表 12 所示的数据。

表 10　加热型烟草制品烟草（基质）中烟碱含量（mg/g）的精密度限值

加热烟草制品	n	$\hat{m}$	重复性限（r）	再现性限（R）
HTP1	10	13	1	5
HTP2	11	12	0	4
HTP3	11	13	1	5

表 11　加热型烟草制品烟草（基质）中甘油含量（mg/g）的精密度限值

加热烟草制品	n	$\hat{m}$	重复性限（r）	再现性限（R）
HTP1	11	133	16	33
HTP2	10	108	11	27
HTP3	11	187	17	52

表 12　加热型烟草制品烟草（基质）中丙二醇含量（mg/g）的精密度限值

加热烟草制品	n	$\hat{m}$	重复性限（r）	再现性限（R）
HTP1	12	3	0	2
HTP2	11	4	0	2
HTP3	9	6	0	1

23　参考文献

23.1　联合国毒品和犯罪问题办公室（UNODC）《代表性药物抽样指南《(维也纳，实验室和科学科，2009 年，http://www.unodc.org/documents/scientific/Drug_Sampling.pdf）。

23.2　WHO TobLabNet SOP 02 烟草制品成分和释放物分析方法验证标准操作规程（世界卫生组织烟草实验室网络，日内瓦，2017 年）。

23.3　ISO 5725-1：检测方法和结果的准确度（正确度和精密度）　第 1 部分 一般原则和定义。

23.4　ISO 5725-2：检测方法和结果的准确度（正确度和精密度）　第 2 部分 确定标准检测方法重复性和再现性的基本方法。

23.5　加热烟草制品（HTP）中烟碱、甘油和丙二醇含量测定分析方法验证的合作研究报告（待出版）。

附录1 加热型烟草制品中烟碱、甘油、丙二醇和三乙酸甘油酯（可选）含量测定的典型色谱图

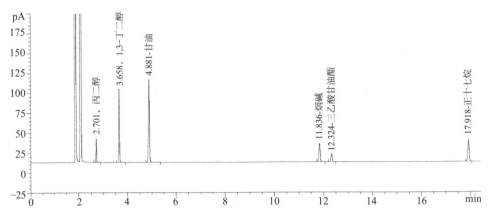

附图1 烟碱浓度为 0.4 mg/mL、甘油浓度为 4.0 mg/mL、丙二醇浓度为 0.5 mg/mL、三乙酸甘油酯浓度为 0.5 mg/mL 的标准溶液的色谱图示例

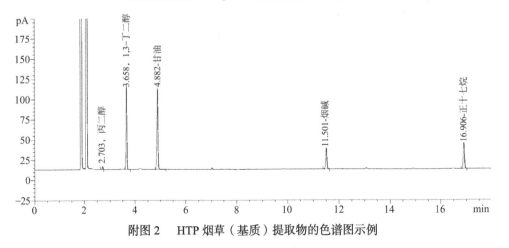

附图2 HTP烟草（基质）提取物的色谱图示例

附录 2　GC-MS 作为替代测量方法的设置示例

特定的 GC 条件，如载气、流速和总运行时间，可根据特定的 MS 需求调整。
MS 操作条件：
传输线温度：≥180℃；
停留时间：50 ms；
电离模式：电子电离（电离电压 70 eV）。
检测：
丙二醇：m/z 61（定量离子）45（定性离子）；
甘油：m/z 61（定量离子）43（定性离子）；
烟碱：m/z 162（定量离子）133（定性离子）；
喹哪啶：m/z 143（定量离子）128（定性离子）；
正十七烷：m/z 240（定量离子）85（定性离子），由于其比质谱较低，不建议使用本内标进行 GC-MS 检测。

推荐的 GC-MS 内标物是分析物的氘化形式。可使用其他内标，如喹哪啶或正十七烷。

使用者应自行验证替代内标物的性能，以确保满足质量控制标准。
为了建立本方法的重复性和再现性数据，在协同实验中不包括 GC-MS 数据。

WHO TobLabNet
Official Method
SOP 11

Standard operating procedure for determination of nicotine, glycerol and propylene glycol in e-liquids

Method:	Determination of nicotine, glycerol and propylene glycol in e-liquids
Analytes:	Nicotine (3-[(2S)-1-methylpyrrolidin-2-yl]pyridine) (CAS # 54-11-5) Glycerol (propane-1,2,3-triol) (CAS # 56-81-5) Propylene glycol (propane-1,2-diol) (CAS # 57-55-6)
Matrix:	e-liquid
Last update:	March 2021

FOREWORD

This document was prepared by members of the World Health Organization (WHO) Tobacco Laboratory Network (TobLabNet) in cooperation with member laboratories of the European Joint Action on Tobacco Control (JATC) as an analytical method standard operating procedure (SOP) for measuring nicotine, glycerol and propylene glycol in e-liquids.

INTRODUCTION

In order to establish comparable measurements for testing e-liquids globally, consensus methods are required for measuring specific contents of e-liquids. The Conference of the Parties (COP) to the WHO Framework Convention on Tobacco Control (WHO FCTC) at its sixth session (Moscow, Russian Federation, 13–18 October 2014) requested the Convention Secretariat to invite WHO to: (a) prepare an expert report on electronic nicotine delivery systems (ENDS) and electronic non-nicotine delivery systems (ENNDS) for the seventh session of the COP (COP7), with an update on the evidence of the health impacts of ENDS/ENNDS, their potential role in quitting tobacco usage and impact on tobacco control efforts; (b) subsequently assess policy options to achieve the objectives outlined in paragraph 2 of decision FCTC/COP6(9); and (c) consider the methods to measure the contents and emissions of these products.[1]

As nicotine content is limited to a certain concentration in some parts of the world (for example, in the European Union, the maximum nicotine concentration in e-liquids is 20 mg/mL), nicotine is considered a priority component to be measured in e-liquids. Since glycerol and propylene glycol are typically ingredients of e-liquids and can be measured simultaneously with nicotine, these components are included in the SOP.

This SOP was prepared to describe the procedure for the determination of nicotine, glycerol and propylene glycol in e-liquid and based on ISO 20714 [**2.1**].

1   SCOPE

This method is suitable for the quantitative determination of nicotine, glycerol and propylene glycol in e-liquids by gas chromatography (GC). The working range of the method is for 1 to 30 mg/mL nicotine, for 200 mg/mL to 1000 mg/mL propylene glycol and for 200 mg/mL to 1000 mg/mL glycerol.

---

[1] Decision FCTC/COP6(9).

2 REFERENCES

2.1 *ISO 20714 (en). E-liquid—Determination of nicotine, propylene glycol and glycerol in liquids used in electronic nicotine delivery devices—Gas chromatographic method (ISO 20714:2019, IDT).*

2.2 *ISO 13276: Tobacco and tobacco products—Determination of nicotine purity— Gravimetric method using tungstosilicic acid.*

2.3 *ISO 5725-2: Accuracy (trueness and precision) or measurement methods and results—Part 2: Basic method for the determination of repeatability and reproducibility of a standard measurement method.*

2.4 *World Health Organization. Standard operating procedure for validation of analytical methods of tobacco product contents and emissions. Geneva, Tobacco Laboratory Network, 2017 (WHO TobLabNet SOP 02) (https://www.who.int/tobacco/publications/prod_regulation/standard-operation-validation-02/en/, accessed 10 December 2020).*

2.5 *United Nations Office on Drugs and Crime. Guidelines on representative drug sampling. Vienna, Laboratory and Scientific Section, 2009 (http://www.unodc.org/documents/scientific/Drug_Sampling.pdf, accessed 10 December 2020).*

3 TERMS AND DEFINITIONS

3.1 *Nicotine content*: total amount of nicotine in e-liquid, expressed as mg per millilitre of e-liquid.

3.2 *E-liquid*: liquid or gel which may or may not contain nicotine intended for aerosolization, to be inhaled with an electronic delivery device.

3.3 *Electronic nicotine delivery device system / electronic non-nicotine delivery system*: device used to aerosolize an e-liquid for inhalation.

3.4 *Laboratory sample*: sample intended for testing in a laboratory, consisting of a single type of product delivered to the laboratory at one time or within a specified period.

3.5 *Test sample*: product to be tested, taken at random from the laboratory sample. The number of products taken shall be representative of the laboratory sample.

3.6 *Test portion*: random portion from the test sample to be used for a single determination. The number of products taken shall be representative of the test sample.

3.7 *Prefilled e-liquid cartridges*: prefilled e-liquid containers that can be or are connected directly to an electronic nicotine delivery device.

4 METHOD SUMMARY

4.1 After bringing to room temperature, the e-liquid is homogenized.

4.2 The e-liquid is diluted with a diluent consisting of propan-2-ol [**7.6**] and internal standards [**7.7**] [**7.8**].

4.3 The ratios of the peak areas of analytes (nicotine, glycerol, propylene glycol) and corresponding internal standards, derived from the measurement of standard solutions with known concentrations, are plotted against the analyte concentration. Calibration curves used to determine the analyte content of each test portion are created by linear regression.

5 SAFETY AND ENVIRONMENTAL PRECAUTIONS

5.1 Follow routine safety and environmental precautions, as in any chemical laboratory activity.

5.2 The testing and evaluation of certain products with this test method may require the use of materials or equipment that could be hazardous or harmful to the environment; this document does not purport to address all the safety aspects associated with its use. All persons using this method have the responsibility to consult with the appropriate authorities and to establish health and safety practices as well as environmental precautions in conjunction with any existing applicable regulatory requirements prior to its use.

5.3 Special care should be taken to avoid inhalation or dermal exposure to harmful chemicals. Use a chemical fume hood, and wear an appropriate laboratory coat, gloves and safety goggles when preparing or handling undiluted materials, standard solutions, diluent solutions or collected samples.

6 APPARATUS AND EQUIPMENT
Usual laboratory apparatus, in particular:

6.1 Positive displacement pipette, applicable for repetitive pipetting 100 µL (e.g. Brand HandyStep Electronic or Eppendorf Repeater stream).

6.2 Positive displacement pipette, applicable for repetitive pipetting 9.9 mL (e.g. Brand HandyStep Electronic or Eppendorf Repeater stream).

6.3 Transparent 20 mL flat bottom sample vials or other suitable flasks.

6.4 3D rotating mixer configured to hold flasks in position (e.g. Stuart Scientific gyro rocker, MiuLab RH-18+ 3D Rotating Mixer).

6.5 Vortex mixer.

6.6 Gas chromatograph equipped with a flame ionization detector (GC-FID).

6.7 Capillary GC column capable of distinct separation of solvent peaks, the peaks for the internal standard, nicotine and other components (e.g. Agilent DB-ALC1 (30 m x 0.32 mm, 1.8 µm)).

6.8 In case of prefilled e-liquid cartridges: centrifuge tubes, with screw cap, designed for at least 1500 rpm (relative centrifugal force (RCF) ≈ 315) and capable of holding different types of prefilled e-liquid cartridges.

7 REAGENTS AND SUPPLIES

All reagents shall be of at least analytical reagent grade unless otherwise noted. When possible, reagents are identified by their Chemical Abstracts Service (CAS) registry numbers.

7.1 Carrier gas: Helium [CAS number: 7440-59-7] of high purity (> 99.999%).

7.2 Auxiliary gases: Air and hydrogen [CAS number: 1333-74-0] of high purity (> 99.999%) for the flame ionization detector.

7.3 Nicotine ((S)-3-[1-Methylpyrrolidin-2-yl]pyridine); –(–)Nicotine [CAS number: 54-11-5] of known purity not less than 98%. Nicotine salicylate [29790-52-1] of known purity not less than 98% may be used.

7.4 Glycerol (propane-1,2,3-triol) [CAS number: 56-81-5] of known purity not less than 98%.

7.5 Propylene glycol (propane-1,2-diol) [CAS number: 57-55-6] of known purity not less than 98%.

7.6 Propan-2-ol [CAS number: 67-63-0], with a maximum water content of 1.0 g/L.

7.7 Internal standard for GC analyses:
Nicotine determination:
n-heptadecane (purity > 98% of mass fraction) [CAS number: 629-78-7]. Quinaldine (purity > 98% of mass fraction) [CAS number: 91-63-4] or other suitable alternatives may be used.

7.8 Internal standard for glycerol and propylene glycol: 1,3-Butanediol, (purity > 99% of mass fraction) [CAS number: 107-88-0].

8 PREPARATION OF GLASSWARE

Clean and dry glassware in a manner to avoid contamination.

9 PREPARATION OF SOLUTIONS

9.1 Diluent solution

The diluent solution consists of propan-2-ol [7.6] containing appropriate amounts of internal standards (0.5 mL/L n-heptadecane / 2 mL/L 1,3-Butanediol).

Pipette 0.50 mL of n-heptadecane [**7.7**] plus 2.0 mL 1,3-Butanediol [**7.8**] into a 1-litre volumetric flask. Dilute to volume (1 litre) with propan-2-ol [**7.6**], mix thoroughly and transfer the solution into a storage container equipped with features to prevent contamination.

Note: The concentration and/or type of internal standard may be adjusted, keeping in mind the possible effect of internal standards on the sensitivity and selectivity, as well as the linear range of the method.

10 PREPARATION OF STANDARDS

Preparation of the standard solutions as described below is for reference purposes. The preparation of the standard solutions can be adjusted if necessary.

10.1 Nicotine standard stock solution (5 g/L)

Weigh approximately 500 mg of nicotine [**7.3**] (or 925 mg nicotine salicylate) to the nearest 0.1 mg into a 100 mL volumetric flask and dilute to volume with the diluent solution [**9.1**].

Mix thoroughly and store between 0 °C and 4 °C and exclude light (maximum storage of four weeks).

Solvent and solutions stored at low temperatures shall be allowed to equilibrate to (22 ± 5) °C before use.

10.2 Glycerol and propylene glycol standard stock solution (50 g/L)

Weigh approximately 5000 mg of glycerol plus 5000 mg of propylene glycol to the nearest 0.1 mg into a 100 mL volumetric flask and dilute to volume with the diluent solution [**9.1**].

Mix thoroughly and store between 0 °C and 4 °C protected from light (maximum storage of four weeks).

Solvent and solutions stored at low temperatures shall be allowed to equilibrate to (22 ± 5) °C before use.

10.3 Working standard solutions

10.3.1 Pipette the designated amount of the nicotine standard stock solution prepared in **10.1** for the specific standard solution into 100 mL volumetric flasks, as described in Table 1.

10.3.2 Add the designated amount of the combined glycerol and propylene glycol standard stock solution prepared in **10.2** into the same 100 mL volumetric flasks [**10.3.1**], as specified in Table 2.

Fill the volumetric flasks to the mark (100 mL) with diluent solution [**9.1**].

Store the standard solutions, protected from light, at 4-8 °C (maximum storage of one week).

10.3.3 Determine the final nicotine, glycerol and propylene glycol concentrations in the calibration standard solutions from:

$$\text{Final concentration (mg/L)} = \frac{x*y}{10000}$$

where x is the original weight (in mg) of the analyte as weighed in **10.1** or **10.2**, and y is the volume of the standard stock solution (in mL) as pipetted in **10.3.1** and **10.3.2**.

The nominal concentrations in the nicotine calibration standard solutions are shown in Table 1, the glycerol and propylene glycol concentrations are shown in Table 2.

Table 1. Concentrations of nicotine in calibration standard solutions

Standard	Volume of nicotine standard stock solution (5 g/L) (mL)	Volume of internal standard solution (µL)	Total volume (mL)	Approximate nicotine concentration in final mixed standard solution (mg/mL)	Corresponding nicotine concentration in e-liquid (mg/mL)
1	0.2		100	0.01	1.0
2	1.0	Not applicable, included in diluent solution	100	0.05	5.0
3	2.0		100	0.10	10.0
4	4.0		100	0.20	20.0
5	6.0		100	0.30	30.0

Table 2. Concentrations of glycerol and propylene glycol in calibration standard solutions

Standard	Volume of glycerol / propylene glycol standard solution (50 g/L) (mL)	Volume of internal standard solution (µL)	Total volume (mL)	Approximate glycerol / propylene glycol concentration in final mixed standard solution (mg/mL)	Corresponding glycerol / propylene glycol concentration in e-liquid (mg/mL)
1	4.0		100	2.0	200
2	8.0	Not applicable, included in diluent solution	100	4.0	400
3	12.0		100	6.0	600
4	16.0		100	8.0	800
5	20.0		100	10.0	1000

The range of the calibration standard solutions may be adjusted, depending on the equipment used and the samples to be tested, keeping in mind the possible effect on the sensitivity of the method.

All solvents and solutions shall be allowed to equilibrate to room temperature (22 ± 5 °C) before use.

11 SAMPLING

11.1 Sample collection

Sampling of e-liquids to obtain a representative sample shall be performed according to applicable legislation or international standards and shall take the availability of samples into account.

11.2 Constitution of test sample

Divide the laboratory sample into separate sales units, if applicable.

For each test sample, take a representative number of products from the laboratory sample.

Each test portion shall consist of at least one unit of the product to be tested.

12 SAMPLE PREPARATION

Due to the high viscosity of e-liquids, special care must be taken when these types of liquids are to be pipetted. For pipetting of high-viscosity liquids, only positive displacement pipettes can be used. Internal laboratory validation showed great effect of pipetting procedures on the results, for this reason a dedicated pipetting procedure is included in the SOP.

If no positive displacement pipettes are available, weighing of the e-liquid can be used as an alternative. However, to be able to report the results in mg/mL units, the density of the e-liquid needs to be determined and the mass of the e-liquid has to be converted into volumetric units (mL).

The specific procedure for pipetting e-liquids is described in **12.1**, extraction of e-liquids from prefilled cartridges is described in **12.2**.

12.1 Refill bottles

12.1.1 Homogenize the e-liquid prior to opening the container. Formation of air bubbles needs to be avoided. If air bubbles are present in the e-liquid, do not use the e-liquid until all air bubbles have disappeared (sonication of the e-liquid will help to remove the air bubbles, avoid increasing temperature caused by sonication).

12.1.2 Aspirate 1 mL of e-liquid, without air bubbles, by using a positive displacement pipette with the possibility of 10-fold repetitive dispensation, set to 100 µL dispensation.

12.1.3 Strike the outer side of the pipette tip against the sample container edge to remove excess liquid.

12.1.4 Dispense twice 100 µL into a waste bin, without touching the waste bin with the pipette tip.

12.1.5 Check if the outside of the pipette tip contains e-liquid remains, if so, repeat step **12.1.3**.

12.1.6 Again, dispense twice 100 μL into a waste bin, without touching the waste bin with the pipette tip.

12.1.7 Check if the outside of the pipette tip contains e-liquid remains.

12.1.8 Pipette 100 μL e-liquid ($V_{e\text{-}liquid}$) into a 20 mL sample vial [**6.3**].

12.1.9 Pipette the remaining e-liquid into a waste bin.

12.1.10 Close the e-liquid container as soon as possible to prevent evaporation or contamination of the sample.

12.2 Prefilled cartridges

To be able to analyse the e-liquid in prefilled cartridges, the liquid needs to be extracted from the cartridge. To avoid contamination of the e-liquid by parts of the cartridge, or tools used for opening the cartridge, a procedure for extracting the total amount of e-liquid from cartridges using a centrifuge is described.

Alternative methods can also be used taking into account that the e-liquid extracted from the cartridge needs to be a representative sample of the cartridge.

12.2.1 Cut off a piece of a disposable pipette tip (100–1000 μL) and place the piece into a suitable centrifuge tube [**6.8**].

12.2.2 Remove the upper cap of the e-liquid cartridge.

12.2.3 Place the cartridge top down into the centrifuge tube, on top of the pipette tip.

12.2.4 Close the centrifuge tube, using the screw cap.

12.2.5 Centrifuge the centrifuge tube for at least five minutes at 1500 rpm (relative centrifugal force (RCF) ≈ 315).

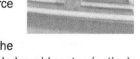

12.2.6 Remove the cartridge and pipette tip from the centrifuge tube (use a pair of tweezers if needed, avoid contamination).

12.2.7 Pipette the e-liquid according to **12.1**.

Note: When less than 1 mL of e-liquid is available for pipetting, an alternative pipetting method needs to be executed.

If for instance only 600 μL e-liquid is available, aspirate 400 μL (instead of 1 mL) and pipette 100 μL e-liquid into the sample vial after the first dispensation into the waste bin.

If less than 500 µL e-liquid is available, aspirate as much e-liquid as possible and pipette 100 µL e-liquid into the sample vial without dispensing e-liquid into the waste bin. If less than 100 µL e-liquid is available, pipette the highest volume possible.

Detail the pipetting procedure in the analysis report if a divergent method (less than 1 mL available) is used.

13 PREPARATION OF THE AEROSOL GENERATING MACHINE
Not applicable for this method.

14 SAMPLE GENERATION
Not applicable for this method.

15 SAMPLE PREPARATION

15.1 Add 9.9 mL diluent solution [9.1] to the 20 mL sample vial containing the 100 µL e-liquid by using the positive displacement pipette [6.2] and close the sample vial.

15.2 Homogenize the liquids in the sample vial by using the vortex mixer [6.5] for at least 30 seconds. Set the speed of the vortex mixer at such a level that a vortex is created during homogenization of the liquid.

15.3 Transfer an aliquot of the homogenized liquid (test solution) into an autosampler vial compatible with the GC-FID instrument.

15.4 If the sample is to be stored, keep it protected from light at 4–8 °C (maximum storage of one week).

Note: Make sure no visual inhomogeneity is present after homogenizing the liquid. If inhomogeneity is still visible, homogenization accordingly [**15.2**] needs to be repeated until no inhomogeneity can be noticed.

16 SAMPLE ANALYSIS

GC with a flame ionization detector is used to quantify nicotine, glycerol and propylene glycol in e-liquids. The analytes are separated from other potential interferences on the column used. Analyte concentrations of test samples are derived by comparison of the area ratio of analyte and internal standard peaks of the test solutions with the area ratio of the analytes in calibration standards with known analyte concentrations.

16.1 GC operating conditions: example

GC column: Agilent DB-ALC1 (30 m x 0.32 mm, 1.8 µm), or equivalent

Column program:
Initial temperature: 140 °C
Hold time: 5:00 min
Temperature rate: 40 °C/min
End temperature: 250 °C
Hold time: 4:00 min
Injection temperature: 225 °C
Detector temperature: 260 °C
Carrier gas: Helium at a flow rate of 1.5 mL/min
Injection volume: 1.0 µL
Injection mode: Split (ratio 1:50)

Note: Adjustment of the operating parameters may be required, depending on the instrument and column conditions as well as the resolution of chromatographic peaks.

Annex 2 provides an example of GC-mass spectrometry (GC-MS) settings to be used if GC-MS is used as an alternative to GC-FID.

16.2 EXPECTED RETENTION TIMES

For the conditions described here, the expected sequence of elution will be propylene glycol, 1,3-Butanediol, glycerol, nicotine and n-heptadecane.

Note 1: Differences in e.g. temperature, gas flow rate and age of the column may alter retention times.

Note 2: The elution order and retention times must be verified before analysis is begun.

Note 3: Under the above conditions, the expected total analysis time will be about 11 min. (The analysis time may be extended to optimize performance.)

16.3 DETERMINATION OF NICOTINE, GLYCEROL AND PROPYLENE GLYCOL

The sequence of measurement of test samples shall be designed in agreement with the applicable laboratory quality system. This section provides an example of a sequence of measurements performed for the determination of nicotine, glycerol and propylene glycol in e-liquids.

Inject aliquots of the calibration standard solutions and test solutions under identical conditions.

16.3.1 Condition the system just before running the sequence by injecting two 1-µL aliquots of a sample test solution as primer.

16.3.2 Inject 1 µL diluent solution [**9.1**] and a test calibration standard solution under the same conditions as the samples to verify the performance of the GC system and component contamination of reagents used.

16.3.3 Inject an aliquot of each solution (nicotine, glycerol and propylene glycol calibration standard solutions) into the GC.

16.3.4 Assess the retention times and responses (area counts) of the standards. If the retention times are similar (± 0.2 min) to the retention times in previous injections, and the responses are within 20% of typical responses in previous injections, the system is ready to perform the analysis. If the retention times and/or responses are outside specifications, seek corrective action according to your laboratory policy.

16.3.5 Record the peak areas of nicotine, glycerol and propylene glycol and the internal standard components.

16.3.6 Calculate the relative response ratios (RF) of the analyte peaks to the internal standard peaks ($RF = A_{analyte} / A_{IS}$) for each analyte of the calibration standard solutions, including the solvent blanks.

16.3.7 Plot the graph of the relative response ratios (Y axis) against the concentration of the analytes (X axis).

16.3.8 Linearity of the calibration curve should cover the entire calibration range.

16.3.9 Perform linear regression ($Y = a + bx$) on these data and use both the slope (b) and the intercept (a) of the linear regression equation for calculation of the results. If the coefficient of determination R^2 is less than 0.99, the calibration should be repeated. Check for individual outliers according to laboratory procedures.

16.3.10 Inject 1 µL of each of the quality control samples and the test samples, and determine the peak areas with the appropriate software.

16.3.11 The signal (peak area) of the analytes obtained for all test solutions must fall within the working range of the calibration curve; otherwise solutions should be diluted with diluent solution (**9.1**) as necessary.

Note: See **Annex 1** for representative chromatograms.

17 DATA ANALYSIS AND CALCULATIONS
17.1 For each test solution, calculate the ratio (Y_t) of the analyte peak area to the internal standard peak area.

17.2 Calculate the analyte concentration (C_{TP} in mg/mL) for each test solution using the coefficients of the linear regression:

$$C_{TP} = \frac{Y_t - a}{b}$$

17.3 Calculate the analyte concentration ($C_{e\text{-liquid}}$) of the e-liquid sample expressed in milligrams per millilitre using the following equation:

$$C_{e\text{-liquid}} = C_{TP} \times \frac{V_{tot}}{V_{e\text{-liquid}}}$$

Where:

$C_{e\text{-liquid}}$ is the concentration of the analyte in the e-liquid, in mg/mL

C_{TP} is the concentration of the analyte in the test solution, in mg/mL

V_{tot} is the total volume of the test solution after dilution of the e-liquid, in mL (default value is 10 mL)

$V_{e\text{-liquid}}$ is the volume of the e-liquid used (**12.1.8**), in mL.

18 SPECIAL PRECAUTIONS

18.1 After installing a new column, condition it by injecting an e-liquid sample solution under the GC conditions described. Injections should be repeated until the peak areas (or heights) of both the component(s) and the internal standard(s) are reproducible. This will require approximately four injections.

18.2 It is recommended to purge high-boiling-point components from the GC column after each sample set (series) by raising the column temperature to 220 °C for 30 min.

18.3 When the peak areas (or heights) for the internal standard(s) are significantly higher than expected, it is recommended to dilute the e-liquid sample without internal standard in the diluent solution and analyse this sample as described in this procedure. This makes it possible to determine whether any component co-elutes with the internal standard, which would cause artificially lower results for the specific analyte(s).

19 DATA REPORTING

19.1 Report individual measurements for each of the samples evaluated.

19.2 Report results as specified by method specifications.

Note: For more information, see WHO TobLabNet SOP 02 [**2.4**].

20 QUALITY CONTROL

20.1 Control parameters

Note 1: If the quality control measurements are outside the tolerance limits of the expected values, appropriate investigation and action must be taken according to laboratory quality procedures.

Note 2: Additional laboratory quality assurance procedures should be carried out in compliance with the policies of the individual laboratory.

20.2 Laboratory reagent blank

To detect potential contamination during sample preparation and analysis, include a procedural blank determination of diluent solution (**9.1**), as described in **16.3.2**. The results should be less than the limit of detection of the respective analyte.

20.3 Quality control sample

To verify the consistency of the entire analytical process, analyse a reference or quality control e-liquid in accordance with the practices of the individual laboratory.

21 METHOD PERFORMANCE SPECIFICATIONS

21.1 Limit of detection (LOD) and limit of quantification (LOQ)

The LOD is defined as the lowest level of the component for which a higher signal is measured than three times the noise of the instrument used and for which the software identifies the component. The LOQ is set to twice the LOD. These are shown in Table 3 (data provided by internal laboratory validation).

Table 3. LOD and LOQ nicotine, glycerol and propylene glycol in e-liquids

Component	LOD (mg/mL)	LOQ (mg/mL)
Nicotine	0.1	0.2
Glycerol	1.0	2.0
Propylene glycol	0.5	1.0

21.2 Laboratory-fortified matrix recovery

The recovery of the analyte(s) spiked into the matrix was used as a surrogate measure of trueness. The recovery of the components was determined by internal laboratory validation as well as in an international collaborative study.

The recovery was determined in an internal laboratory validation by preparing recovery samples by weighing different amounts of nicotine, glycerol and propylene glycol into 1L flasks. After homogenizing the recovery samples for two hours using a 3D rotating mixer [**6.4**], the recovery samples were transferred into 2 mL vials. Nicotine, glycerol and propylene glycol were determined in each of the recovery samples, analysing the individual vials in five-fold on one day. The native (not spiked) e-liquid was also analysed. The recovery is calculated from the following formula and is summarized for nicotine in Table 4 and for glycerol and propylene glycol in Table 5.

$$Recovery\ (\%) = 100 \times \frac{(\text{concentration analyte recovery sample} - \text{concentration analyte native sample})}{\text{nominal analyte concentration recovery sample}}$$

Table 4. Mean and recovery of nicotine by internal laboratory validation

Spiked amount (mg/mL)	Nicotine	
	Mean (mg/mL)	Recovery (%)
0.195	0.235	120.6
3.098	3.117	96.5
12.388	12.410	101.1
20.209	20.232	98.9

Table 5. Mean and recovery of glycerol and propylene glycol by internal laboratory validation

Glycerol			Propylene glycol		
Spiked amount (mg/mL)	Mean (mg/mL)	Recovery (%)	Spiked amount (mg/mL)	Mean (mg/mL)	Recovery (%)
378.7	377.7	99.7	727.4	723.1	99.4
630.5	636.5	101.0	519.6	515.1	99.1
883.0	898.2	101.7	311.2	307.3	98.7

The recovery was determined in an international collaborative study by weighing four different amounts of nicotine, glycerol and propylene glycol into flasks. After homogenizing the flasks using a 3D rotating mixer [**6.4**], the recovery samples are transferred into 10 mL flasks. For each of the four spiked samples nicotine, glycerol and propylene glycol are determined by analysing the individual flasks in two-fold in an international collaborative study (conducted in 2020). The recovery is calculated from the following formula and is summarized for nicotine in Table 6 and for glycerol and propylene glycol in Table 7.

$$\text{Recovery (\%)} = 100 \times \frac{(\text{concentration analyte recovery sample} - \text{concentration analyte native sample})}{\text{nominal analyte concentration recovery sample}}$$

Table 6. Mean and recovery of nicotine obtained in an international collaborative study

Sample	Nicotine		
	Theoretical concentration (mg/mL)	Mean value study (mg/mL)	Recovery (%)
A	0.25	0.326	130.4%
B	5.08	5.08	100.0%
C	8.03	7.91	98.5%
D	21.27	22.06	103.7%

Table 7. Mean and recovery of glycerol and propylene glycol obtained in an international collaborative study

Sample	Glycerol (mg/mL)			Propylene glycol (mg/mL)		
	Theoretical concentration (mg/mL)	Mean value study (mg/mL)	Recovery (%)	Theoretical concentration (mg/mL)	Mean value study (mg/mL)	Recovery (%)
A	568.0	567.5	99.9%	568.7	563.1	99.0%
B	213.9	211.5	98.9%	855.0	843.2	98.6%
C	772.5	738.0	95.5%	278.3	268.4	96.4%
D	321.5	316.7	98.5%	749.9	736.8	98.2%

21.3 Analytical specificity

The retention time of the analyte of interest is used to verify the analytical specificity. An established range of ratios of the response of the analyte to that of the internal standard of quality control e-liquid is used to verify the specificity of the results for an unknown sample.

21.4 Linearity

The established nicotine calibration curves are linear over the standard concentration range of 0.01–0.30 mg/mL (1.0–30 mg/mL e-liquid).

The propylene glycol and glycerol calibration curves established are linear over the standard concentration range of 2.0–10.0 mg/mL (200–1000 mg/mL e-liquid).

21.5 Possible interference

The presence of flavourings can cause interference, due to a similar retention time to one of the components or internal standard components.

22 REPEATABILITY AND REPRODUCIBILITY

An international collaborative study [**23.4**] conducted in 2020, involving 23 laboratories and five samples (four spiked e-liquids and one commercial e-liquid) gave the following values for this method.

The difference between two single results found on matched e-liquid samples by the same operator using the same apparatus within the shortest feasible time will exceed the repeatability limit, r, on average not more than once in 20 cases with normal, correct application of the method.

Single results for matched e-liquid samples reported by two laboratories will differ by more than the reproducibility limit, R, on average no more than once in 20 cases with normal, correct application of the method.

The test results were analysed statistically in accordance with ISO 5725-1 [**23.1**] and ISO 5725-2 [**2.3**] to give the precision data shown in Tables 8–10.

Table 8. Precision limits for determination of nicotine (mg/mL) in e-liquids

E-liquid	n	$\hat{m}$	Repeatability$_{limit}$	Reproducibility$_{limit}$
Sample A	15	0.326	0.079	0.277
Sample B	22	5.08	0.52	1.06
Sample C	21	7.91	0.26	1.49
Sample D	21	22.06	0.66	3.74
Sample E	22	11.38	1.30	1.93

n	is the number of laboratories that participated
$\hat{m}$	is the mean value of nicotine content in e-liquid
Repeatability$_{limit}$	is the repeatability limit of nicotine content in e-liquid
Reproducibility$_{limit}$	is the reproducibility limit of nicotine content in e-liquid

Table 9. Precision limits for determination of glycerol (mg/mL) in e-liquids

E-liquid	n	$\hat{m}$	Repeatability limit	Reproducibility limit
Sample A	20	567.5	13.5	48.1
Sample B	21	211.5	8.5	43.6
Sample C	21	738.0	49.3	69.9
Sample D	20	316.7	17.2	32.1
Sample E	21	365.5	16.1	40.7

n	is the number of laboratories that participated
$\hat{m}$	is the mean value of glycerol content in e-liquid
Repeatability$_{limit}$	is the repeatability limit of glycerol content in e-liquid
Reproducibility$_{limit}$	is the reproducibility limit of glycerol content in e-liquid

Table 10. Precision limits for determination of propylene glycol (mg/mL) in e-liquids

E-liquid	n	$\hat{m}$	Repeatability limit	Reproducibility limit
Sample A	22	563.1	13.1	23.2
Sample B	22	843.2	37.6	60.7
Sample C	22	268.4	15.5	18.3
Sample D	22	736.8	21.5	38.6
Sample E	23	570.1	58.3	58.3

n	is the number of laboratories that participated
$\hat{m}$	is the mean value of propylene glycol content in e-liquid
Repeatability$_{limit}$	is the repeatability limit of propylene glycol content in e-liquid
Reproducibility$_{limit}$	is the reproducibility limit of propylene glycol content in e-liquid

23 BIBLIOGRAPHY

23.1 ISO5725-1. Accuracy (trueness and precision) of measurement methods and results – Part 1: General principles and definitions.

23.2 ISO 5725-4. Accuracy (trueness and precision) of measurement methods and results – Part 4: Basic methods for the determination of the trueness of a standard measurement method.

23.3 ISO Standards – Products by TC: ISO/TC 126 (http://www.iso.org/iso/home/store/catalogue_tc/catalogue_tc_browse.htm?commid=52158).

23.4 Report of a collaborative study for the validation of an analytical method for the determination of nicotine, glycerol and propylene glycol in e-liquids (in press).

ANNEX 1

Typical chromatograms obtained in the analysis of e-liquid for nicotine, propylene glycol and glycerol content

Fig. A1.1. Example of a chromatogram of a standard solution with a nicotine concentration of 0.30 mg/mL and a propylene glycol and glycerol concentration of 10.0 mg/mL

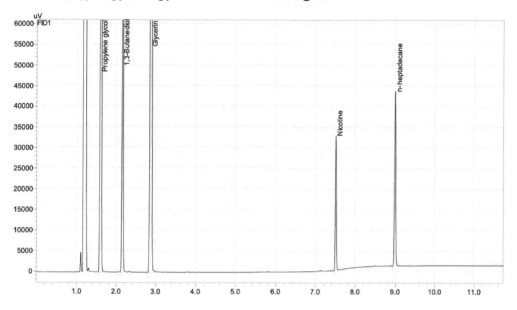

Fig. A2.2. Example of a chromatogram of a sample solution

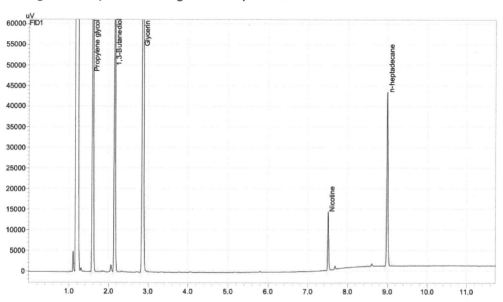

ANNEX 2

GC-MS settings for alternative measurement technique

Mass spectrometry (MS) operating conditions

Specific gas chromatography (GC) conditions, like carrier gas, flow rate and total run time, can be adapted to specific MS needs.

Transfer line temperature: >= 180°C

Dwell time:	50 msec
Ionization mode:	Electron Ionization (electron energy 70 eV)
Detection:	propylene glycol: m/z 61 (quantifier) 45 (qualifier) glycerol: m/z 61 (quantifier ion) 43 (confirmation ion) nicotine: m/z 162 (quantifier ion) 133 (confirmation ion) quinaldine: m/z 143 (quantifier ion) 128 (confirmation ion)
	n-heptadecane: m/z 240 (quantifier ion) 85 (confirmation ion). Use of this internal standard is not recommended for GC-MS detection due to its low specific mass spectrum.

WHO TobLabNet
Official Method
SOP 12

Standard operating procedure for determination of nicotine content in smokeless tobacco products

Method:	Determination of nicotine content in smokeless tobacco products
Analytes:	Nicotine (3-[(2S)-1-methylpyrrolidin-2-yl]pyridine) (CAS # 54-11-5)
Matrix:	Smokeless tobacco products
Last update:	December 2021

FOREWORD

Smokeless tobacco products are gradually attracting the interest of public health organizations. A request was made by the WHO Framework Convention on Tobacco Control (WHO FCTC) Conference of the Parties (COP) at its fifth session (Seoul, 2012) to identify options to regulate chemicals in smokeless tobacco products. This document is prepared in response to the request made by the COP at its seventh session (Delhi, 2016) to the WHO FCTC Secretariat to invite WHO to finalize the standard operating procedures (SOPs) for measuring nicotine and tobacco-specific nitrosamines as requested by decision FCTC/COP6(12) 2b.ii. In pursuance of this request, WHO organized a collaborative study involving its Tobacco Laboratory Network (TobLabNet) testing laboratories, which tested materials for which there was some chemical characterization, represented a range of common forms of smokeless tobacco products and differed in physical and chemical properties. The assessment of the applicability and adaptability of validated WHO SOPs to smokeless tobacco products and the recommended methods are presented in this SOP.

This document was prepared by members of the WHO TobLabNet as an analytical method SOP for measuring nicotine in smokeless tobacco products.

INTRODUCTION

In order to establish comparable measurements for testing tobacco products globally, consensus methods are required to measure specific parameters. WHO TobLabNet reviewed commonly used procedures for the determination of nicotine in smokeless tobacco products in order to prepare a procedure for a WHO TobLabNet SOP.

This SOP was adapted from WHO TobLabNet SOP 04 [**2.1**] to describe the procedure for determination of nicotine in smokeless tobacco products.

1.   SCOPE

This method is suitable for quantitative determination of nicotine in smokeless tobacco products by gas chromatography (GC)–flame ionization detection.

2.   REFERENCES

2.1   World Health Organization. Standard operating procedure for determination of nicotine in cigarette tobacco filler. Geneva, Tobacco Laboratory Network, 2017 (WHO TobLabNet SOP 04).

2.2   United Nations Office on Drugs and Crime. Guidelines on representative drug sampling. Vienna, Laboratory and Scientific Section, 2009 (http://www.unodc.org/documents/scientific/Drug_Sampling.pdf).

2.3 World Health Organization. Standard operating procedure for validation of analytical methods of tobacco product contents and emissions. Geneva, Tobacco Laboratory Network, 2017 (WHO TobLabNet SOP 02).

2.4 ISO5725-1. Accuracy (trueness and precision) of measurement methods and results – Part 1: General principles and definitions.

2.5 ISO 5725-2: Accuracy (trueness and precision) or measurement methods and results – Part 2: Basic method for the determination of repeatability and reproducibility of a standard measurement method.

3 TERMS AND DEFINITIONS

3.1 *Nicotine content*: Total amount of nicotine in smokeless tobacco products, expressed as milligrams per gram dry weight.

3.2 *Smokeless tobacco*: Tobacco-containing part of a smokeless tobacco product.

3.3 Smokeless *tobacco products*: Products made entirely or partly of leaf tobacco as the raw material that are manufactured to be used by sucking, chewing or snuffing (Article 1(f) of the WHO FCTC), including snus (dry and wet), chewing tobacco or a mixture of material originating from a tobacco plant.

3.4 *Laboratory sample*: Sample intended for testing in a laboratory, consisting of a single type of product delivered to the laboratory at one time or within a specified period.

3.5 *Test sample*: Product to be tested, taken at random from the laboratory sample. The number of products taken shall be representative of the laboratory sample.

3.6 *Test portion:* Random portion of the test sample to be used for a single determination. The number of products taken shall be representative of the test sample.

4. METHOD SUMMARY

4.1 Nicotine is extracted from the smokeless tobacco with a mixture of *n*-hexane, sodium hydroxide solution and water.

4.2 The organic layer is analysed by GC with a flame ionization detector.

4.3 The ratio of nicotine peak area to internal standard is compared on a calibration curve created by analysis of standards with known concentrations of nicotine to determine the nicotine content of each test portion.

5. SAFETY AND ENVIRONMENTAL PRECAUTIONS

5.1 Take routine safety and environmental precautions, as in any chemical laboratory activity.

5.2 The testing and evaluation of certain products with this test method may require the use of materials or equipment that could be hazardous or harmful to the environment; this document does not purport to address all safety aspects associated with its use. All persons using this method have the responsibility to consult the appropriate authorities and to establish health and safety practices as well as environmental precautions in conjunction with any existing applicable regulatory requirements prior to its use.

5.3 Special care should be taken to avoid inhalation or dermal exposure to harmful chemicals. Use a chemical fume hood, and wear an appropriate laboratory coat, gloves and safety goggles when preparing or handling undiluted materials, standard solutions, extraction solutions or collected samples.

6. APPARATUS AND EQUIPMENT
Usual laboratory apparatus, in particular:

6.1 Extraction vessels: Erlenmeyer flasks (100 mL) with stoppers, 100 mL Pyrex bottles with crimp seals and septa, 100 mL culture tubes with Teflon-lined caps or other suitable flasks.

6.2 Shaker (linear type) configured to hold the extraction vessels in position.

6.3 Capillary GC equipped with a flame ionization detector.

6.4 Capillary GC column capable of distinct separation of peaks for the solvent, the internal standard, nicotine and other tobacco components (e.g. Varian WCOT Fused Silica, 25 m × 0.25 mm ID; coating: CP-WAX 51).

6.5 Ultrasonic bath.

7. REAGENTS AND SUPPLIES
All reagents shall be of at least analytical reagent grade unless otherwise noted. When possible, reagents are identified by their Chemical Abstract Service (CAS) registry numbers.

7.1 Carrier gas: Helium [7440-59-7] of high purity (> 99.999%).

7.2 Auxiliary gases: Air and hydrogen [1333-74-0] of high purity (> 99.999%) for the flame ionization detector.

7.3 *n*-Hexane [110-54-3], GC grade, with a maximum water content of 1.0 g/L.

7.4 –(–)Nicotine [54-11-5] of known purity not less than 98%. Nicotine salicylate [29790-52-1] of known purity not less than 98% may be used.

7.5 Sodium hydroxide [1310-73-2] pellets.

7.6 Internal standard: *n*-heptadecane (purity > 98% of mass fraction) [629-78-7]. Quinaldine [91-63-4], isoquinoline [119-65-3], quinoline [91-22-5] or other suitable alternatives may be used.

8. PREPARATION OF GLASSWARE
8.1 Clean and dry glassware in a manner to avoid contamination.

9. PREPARATION OF SOLUTIONS
9.1 Sodium hydroxide solution (2 mol/L)

9.1.1 Weigh approximately 80 g of sodium hydroxide.

9.1.2 Dissolve measured sodium hydroxide in water, and dilute with water to 1 L.

9.2 Extraction solution (0.5 mg/mL)

9.2.1 Weigh approximately 0.5 g (to 0.001 g accuracy) of *n*-heptadecane or alternative internal standard.

9.2.2 Dissolve measured *n*-heptadecane or alternative internal standard in *n*-hexane, and dilute to 1 L with *n*-hexane.

10. PREPARATION OF STANDARDS
The method for preparing standard solutions described below is for reference purposes and can be adjusted if necessary.

10.1 Nicotine standard stock solution (2 g/L)

10.1.1 Weigh approximately 200 mg nicotine or 370 mg nicotine salicylate to 0.0001 g accuracy into a 200-mL (or 250-mL) Erlenmeyer flask.

10.1.2 Dissolve the measured nicotine in 50 mL of water.

10.1.3 Pipette 100 mL of extraction solution (**9.2.2**) and add 25 mL of 2 mol/L sodium hydroxide solution.

10.1.4 Shake the two-phase mixture obtained vigorously for 60 ± 2 min in a shaker. Take care to mix the phases well.

10.1.5 Draw out the supernatant organic phase for standard solution preparation. If necessary, store this solution, protected from light, at 4–8 °C.

10.2 Nicotine standard solutions

10.2.1 Prepare the calibration solutions using the standard stock solution prepared in **10.1.4**, according to the scheme given in Table 1.

10.2.2 Fill the volumetric flasks to the mark with extraction solution (**9.2.2**).

10.2.3 The standard solutions may be stored at 4–8 °C, protected from light.

10.2.4 Determine the final nicotine concentrations in the standard solutions from:

$$\text{Final concentration (mg/L)} = x * y * \frac{1000}{100 * 20}$$

Where x is the original weight (in mg) of nicotine as weighed in **10.1.1**, and y is the volume of the stock standard solution as pipetted in **10.2.1**.

The final nicotine concentrations in the standard solutions are shown in Table 1.

Table 1. Concentrations of nicotine in standard solutions

Standard	Volume of nicotine standard solution (2 g/L) (mL) (y)	Volume of internal standard solution (μL)	Total volume (mL)	Nicotine concentration in final mixed standard solution (mg/L)	Approximate level equivalent to unknown levels in smokeless tobacco (mg/g) when 1.5 g of sample taken
1	0.5	Not applicable, included in extraction solution	20	50	1.3
2	2.5		20	250	6.7
3	5.0		20	500	13.3
4	7.5		20	750	20.0
5	10.0		20	1000	26.7
6	15.0		20	1500	40.0

The range of the standard solutions may be adjusted, within the calibration range, depending on the equipment used and the samples to be tested, keeping in mind the possible effect on the sensitivity of the method.

All solvents and solutions must be adjusted to room temperature before use.

11. SAMPLING

11.1 Sample smokeless tobacco product according to laboratory sampling procedure. Alternative approaches may be used to obtain a representative laboratory sample in accordance with individual laboratory practice or when required by specific regulation or availability of samples.

11.2 Constitution of test sample

11.2.1 Divide the laboratory sample into separate units (e.g. packet, container), if applicable.

11.2.2 Take an equal amount of products for each test sample from at least $\sqrt{n}$[**2.2**] of the individual units (e.g. packet, container).

12. PRODUCT PREPARATION

12.1 Remove the smokeless tobacco product from the pack or container. Include quality control samples (when applicable).

12.2 Take an appropriate representative portion of the smokeless tobacco product according to individual laboratory practice (e.g., food analysis sampling approach may be applied).

12.3 Extract the smokeless tobacco from the product.

12.4 Combine and mix sufficient amounts of smokeless tobacco product samples to constitute about 0.5–2 g of homogeneous smokeless tobacco for each test sample.

13. PREPARATION OF THE SMOKING MACHINE
Not applicable

14. SAMPLE GENERATION
Not applicable

15. SAMPLE PREPARATION
15.1 Take 0.5–2 g of the sample and weigh it to 0.001-g accuracy into a 100-mL extraction vessel.

15.2 Mix the test sample with 20 mL of water, 40 mL of extraction solution (**9.2.2**) and 10 mL of 2 mol/L sodium hydroxide solution.

15.3 Shake the flask for 60 ± 2 min on a shaker.

15.4 Leave the sample flask to stand for another 20 min to allow visible, clear separation of the phases. After separation of the phases, analyse an aliquot of the organic (upper) phase as rapidly as possible by GC. If the phases do not separate clearly, place the Erlenmeyer flask in an ultrasonic bath until the phases are clearly separated.

15.5 Additional steps if solutions are found to be cloudy or murky:
(Option 1) To filter the sample through Whatman paper No. 41.
(Option 2) To centrifuge samples at 10 000 rpm or at a suitable relative centrifugal force (> 500) to ensure clear solutions.
(RCF > 500) to ensure clear solutions are obtained

15.6 If the sample is to be stored, keep it protected from light at 4–8 °C.

16. SAMPLE ANALYSIS
GC coupled with a flame ionization detector is used to quantify nicotine in smokeless tobacco products. The analytes are resolved from other potential interference on the GC column. Comparison of the area ratio of the unknowns with the area ratio of the known standard concentrations yields individual analyte concentrations.

16.1 GC operating conditions

GC column:	Varian WCOT fused silica, 25 m × 0.25 mm ID
Coating:	CP-WAX 51 or equivalent
Column temperature:	170 °C (isothermal)
Injection temperature:	270 °C
Detector temperature:	270 °C
Carrier gas:	Helium at a flow rate of 1.5 mL/min
Injection volume:	1.0 µL
Injection mode:	Split 1:10

Note: The operating parameters might have to be adjusted, depending on the instrument and column conditions and the resolution of chromatographic peaks.

16.2 Expected retention times

16.2.1 For the conditions described here, the expected sequence of elution will be *n*-heptadecane, nicotine.

16.2.2 Differences in, e.g., temperature, gas flow rate and age of the column, may alter retention times.

16.2.3 The elution order and retention times must be verified before the analysis of samples.

16.2.4 Under the above conditions, the expected total analysis time will be about 6 min. The analysis time may be extended to optimize performance.

16.3 Determination of nicotine

The sequence of determination will be in accordance with individual laboratory practice. This section gives an example of a sequence of operations for the determination of nicotine in smokeless tobacco products.

16.3.1 Inject an aliquot of n-hexane (7.3) to check for any contamination in the system or reagents.

16.3.2 Condition the system just before use by injecting two 1-µL aliquots of a sample solution as a primer.

16.3.3 Inject 1 µL extraction solution (9.2) and a test calibration standard solution under the same conditions as the samples to verify the performance of the GC system.

16.3.4 Inject 1 µL n-hexane (7.3) to check for any contamination of the system or reagents.

16.3.5 Inject in random order an aliquot of each nicotine standard solution into the GC.

16.3.6 Assess the retention times and responses (area counts) of the standards. If the retention times are similar (± 0.2 min) to the retention times in previous injections and the responses are within 20% of typical responses in previous injections, the system is ready to perform the analysis. If the responses are outside specifications, seek corrective action according to your laboratory policy.

16.3.7 Record the peak areas of nicotine and the internal standard.

16.3.8 Calculate the relative response ratio (*RF*) of the nicotine peak to the internal standard peak ($RF = A_{nicotine} / A_{IS}$) for each of the nicotine standard solutions, including the solvent blanks.

16.3.9 Plot a graph of the concentration of nicotine (X axis) against the area ratios (Y axis).

16.3.10 The intercept should not be statistically significantly different from zero.

16.3.11 The standard curve shall be linear over the entire calibration range.

16.3.12 Calculate the calibration curve ($Y = a + bx$) by linear regression from these data and use both the slope (*b*) and the intercept (*a*) of the calibration curve for calculation of the results. If the coefficient of determination R^2 is < 0.99, the calibration should be repeated. If an individual calibration point differs by more than 10% from the calculated value (estimated from the calibration curve), the calibration point should be omitted.

16.3.13 Inject 1 µL of each of the quality control sample (**20.3**) and the test sample extracts (**15.4**), and determine the peak areas with the appropriate software.

16.3.14 The signal (peak area ratio) obtained for all test portions must fall within the working range of the calibration curve; otherwise, the test portion size must be adjusted or the test sample extract must be diluted.

See Annex 1 for representative chromatograms.

17 DATA ANALYSIS AND CALCULATIONS

17.1 For each test portion, calculate the ratio (Y_t) of the nicotine peak area to the internal standard peak area.

17.2 Calculate the nicotine concentration in mg/L for each test portion aliquot using the coefficients of the calibration curve ($m_t = (Y_t - a) / b$).

17.3 Calculate the nicotine content, m_n, of the tobacco sample expressed in milligrams per gram from the following equation:

$$m_n = \frac{m_t * V_e}{m_o * 1000}$$

where m_t is the concentration of nicotine in the test solution, in mg/L; V_e is the volume of the extraction solution used, in mL; and m_o is the mass of the test portion, in g.

18. SPECIAL PRECAUTIONS

18.1 After installing a new column, condition it by injecting a tobacco sample extract under the GC conditions described above. Injections should be repeated until the peak areas (or heights) of both the nicotine and the internal standard are reproducible. This will require approximately four injections.

18.2 It is recommended that high-boiling-point components be purged from the GC column after each sample set (series) by raising the column temperature to 220 °C for 30 min.

18.3 When the peak areas (or heights) for the internal standard are significantly higher than expected, it is recommended that the tobacco sample be extracted without internal standard in the extraction solution. This makes it possible to determine whether any component co-elutes with the internal standard, which would cause artificially lower content values for nicotine.

19. DATA REPORTING

19.1 Report individual measurements for each sample evaluated.

19.2 Report results as specified in the overall project specifications.

19.3 For more information, see WHO TobLabNet SOP 02 [**2.3**].

20. QUALITY CONTROL

20.1 Control parameters

Note: If the control measurements are outside the tolerance limits of the expected values, appropriate investigation and action must be taken.

Note: Additional laboratory quality assurance procedures should be carried out if necessary, in order to comply with the policies of individual laboratories.

20.2 Laboratory reagent blank

To detect potential contamination during sample preparation and analysis, include a laboratory reagent blank, as described in **16.3.4**. The blank consists of all reagents and materials used in analysing test samples and is analysed in the same way as a test sample. The content of the blank should be below the limit of detection.

20.3 Quality control sample

To verify the consistency of the entire analytical process, analyse a reference tobacco product such as Coresta Reference Products (CRP), when available, in accordance with the practices of individual laboratories.

21. METHOD PERFORMANCE SPECIFICATIONS

21.1 Limit of reporting

The limit of reporting is set to the lowest concentration of the calibration standards used, recalculated to mg/g (e.g., 1.3 mg/g with 50 mg/L as the lowest calibration standard concentration).

21.2 Internal quality control

Recovery of reference material is a surrogate measure of accuracy. Recovery is determined by measuring the level of nicotine in reference smokeless tobacco products. The recovery is calculated from the following equation.

Recovery (%) = 100 × (analytical result / certified amount)

Table 2. Mean and recovery of nicotine content in smokeless tobacco products

Smokeless tobacco sample	Certified value (mg/g)	Mean nicotine content (mg/g)	Recovery (%)
CRP 1	8.0	7.40	92.4
CRP 2	12.0	10.58	88.2
CRP 3	17.0	16.54	97.3
CRP 4	9.0	9.03	100.4

21.3 Analytical specificity

The retention time of the analyte of interest is used to verify analytical specificity. An established range of ratios of the response of the component to that of the internal standard component of a quality control smokeless tobacco product is used to verify the specificity of the results for an unknown sample.

21.4 Linearity

The nicotine calibration curves established are linear over the standard concentration range of 50–1500 mg/L (1.3–40 mg/g).

21.5 Possible interference

The presence of eugenol or flavours can cause interference, as their retention times are similar to that of nicotine. This interference is most likely to occur with samples containing clove or added flavours. The laboratory can resolve the interference by adjusting the chromatographic parameters.

22. REPEATABILITY AND REPRODUCIBILITY

An international collaborative study conducted between September 2020 and March 2021, involving 13 laboratories and four samples (four CRP smokeless tobacco products), performed according to WHO TobLabNet Method Validation Protocol and this SOP gave the following values for this method.

The test results were analysed statistically in accordance with ISO 5725-1 [**2.4**] and ISO 5725-2 [**2.5**] to give the precision data shown in Table 3.

Table 3. Precision limits for determination of nicotine content (mg/g) in smokeless tobacco products

Reference tobacco product	n	$\hat{m}$	Repeatability limit (r)	Reproducibility limit (R)
CRP1	9	7.40	0.49	2.92
CRP2	8	10.58	0.38	2.75
CRP3	7	16.54	0.50	2.77
CRP4	10	9.03	0.49	1.90

APPENDIX 1.

Typical chromatogram obtained in the analysis of smokeless tobacco products for nicotine content

Fig. 1. Example of a chromatogram of a standard solution with a nicotine concentration of 250mg/L

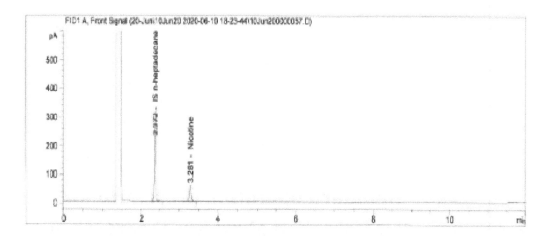

WHO TobLabNet
Official Method
SOP 13

Standard operating procedure for determination of moisture content in smokeless tobacco products

Method:	Determination of moisture content in smokeless tobacco products
Matrix:	Smokeless tobacco products
Last update:	December 2021

FOREWORD

Smokeless tobacco products are gradually attracting the interest of public health organizations. A request was made by the WHO Framework Convention on Tobacco Control (WHO FCTC) Conference of the Parties (COP) at its fifth session (Seoul, 2012) to identify options to regulate chemicals in smokeless tobacco products. This document is prepared in response to the request made by the COP at its seventh session (Delhi, 2016) to the WHO FCTC Secretariat to invite WHO to finalize the standard operating procedures (SOPs) for measuring nicotine and tobacco-specific nitrosamines, as requested by decision FCTC/COP6(12) 2b.ii. In pursuance of this request, WHO organized a collaborative study involving its Tobacco Laboratory Network (TobLabNet) testing laboratories, which tested materials for which some chemical characterization was available, represented a range of common forms of smokeless tobacco product and differed in physical and chemical properties. The assessment of applicability and adaptability of validated WHO SOPs to smokeless tobacco products and the recommended methods are presented in this SOP.

This document was prepared by members of the WHO TobLabNet as an analytical method SOP for measuring the moisture content of smokeless tobacco products. Moisture content is one of the key influences on nicotine delivery by a product.

INTRODUCTION

In order to establish comparable measurements of the moisture content of smokeless tobacco products and to prepare a procedure for WHO TobLabNet products globally, consensus methods are required to measure specific parameters in smokeless tobacco products. The WHO TobLabNet reviewed commonly used procedures in developing this SOP, such as CORESTA Recommended Method No. 76 [**2.1**] and AOAC Official method 966.02 [**2.2**].

1. SCOPE

This method specifies an oven-drying method for determining the moisture content of smokeless tobacco products. Moisture content (oven volatiles) is the reduction of the mass on drying a sample in a forced draft oven at a temperature regulated at 100 °C ± 1 °C for 3 h ± 0.5 min. This method allows measurement of the volatile constituents of smokeless tobacco products, including water and flavouring components that are lost under the specified conditions. Accurate determination of moisture content in smokeless tobacco products is critical, as moisture content affects product stability and product integrity.

2. REFERENCES

2.1 CORESTA Recommended Method No. 76 Determination of Moisture Content (Oven Volatiles) of Tobacco and Tobacco Products

2.2 AOAC Official Method 966.02 Loss on Drying (Moisture) in Tobacco Gravimetric Method

2.3 United Nations Office on Drugs and Crime. Guidelines on representative drug sampling. Vienna, Laboratory and Scientific Section, 2009 (http://www.unodc.org/documents/scientific/Drug_Sampling.pdf).

2.4 World Health Organization. Standard operating procedure for validation of analytical methods of tobacco product contents and emissions. Geneva, Tobacco Laboratory Network, (WHO TobLabNet SOP 02).

2.5 ISO5725-1. Accuracy (trueness and precision) of measurement methods and results – Part 1: General principles and definitions.

2.6 ISO 5725-2: Accuracy (trueness and precision) or measurement methods and results – Part 2: Basic method for the determination of repeatability and reproducibility of a standard measurement method.

3. TERMS AND DEFINITIONS

3.1 *Moisture content*: Moisture content in smokeless tobacco products, expressed as milligrams per gram.

3.2 *Smokeless tobacco*: Tobacco-containing part of a smokeless tobacco product.

3.3 *Smokeless tobacco products*: Products made entirely or partly of leaf tobacco as the raw material that are manufactured to be used by smoking, sucking, chewing or snuffing (Article 1(f) of the WHO FCTC), including snus (dry and wet), chewing tobacco or a mixture of material originating from a tobacco plant.

3.4 *Laboratory sample*: Sample intended for testing in a laboratory, consisting of a single type of product delivered to the laboratory at one time or within a specified period.

3.5 *Test sample*: Product to be tested, taken at random from the laboratory sample. The number of products taken shall be representative of the laboratory sample.

3.6 *Test portion:* Random portion from the test sample to be used for a single determination. The number of products taken shall be representative of the test sample.

3.7 *Wet smokeless tobacco weight:* Weight of the smokeless tobacco product before placement in the oven for testing its moisture content.

3.8 *Dry smokeless tobacco weight*: Weight of the smokeless tobacco product after placement in the oven for testing its moisture content.

4. METHOD SUMMARY

4.1 Moisture content is defined in this method as the reduction in mass when a sample is dried in an air draft oven at a temperature of 100 °C ± 1 °C for 3 h ± 0.5 min.

4.2 After drying, the samples are cooled in a desiccator to room temperature to prevent uptake of humidity from the air.

5. SAFETY AND ENVIRONMENTAL PRECAUTIONS

5.1 Take routine safety and environmental precautions, as in any chemical laboratory activity.

5.2 The testing and evaluation of certain products with this test method may require the use of materials or equipment that could be hazardous or harmful to the environment; this document does not purport to address all the safety aspects associated with its use. All persons using this method have the responsibility to consult the appropriate authorities and to establish health and safety practices as well as environmental precautions in conjunction with any existing applicable regulatory requirements prior to its use.

5.3 Special care should be taken to avoid inhalation or dermal exposure to harmful chemicals. Use a chemical fume hood, and wear an appropriate laboratory coat, gloves and safety goggles when preparing or handling undiluted materials, standard solutions, extraction solutions or collected samples.

6. APPARATUS AND EQUIPMENT
Usual laboratory apparatus, in particular:

6.1 Air draft oven capable of maintaining the air temperature at 100 °C ± 1 °C.

6.2 Desiccator.

6.3 Analytical balance.

6.4 Crucible or evaporating dish to contain the samples, or equivalent.

7. REAGENTS AND SUPPLIES

7.1 Ultrapure water.

8. PREPARATION OF GLASSWARE

8.1 Clean and dry glassware in a manner to avoid contamination.

9. PREPARATION OF SOLUTIONS
Not applicable

10. PREPARATION OF STANDARDS
Not applicable

11. SAMPLING

11.1 Sample smokeless tobacco products according to the laboratory sampling procedure. Alternative approaches may be used to obtain a representative laboratory sample in accordance with individual laboratory practice or when required by specific regulation or availability of samples.

11.2 Constitution of test sample

11.2.1 Divide the laboratory sample into separate units (e.g., packet, container), if applicable.

11.2.2 Take an equal amount of products for each test sample from at least $\sqrt{n}$[**2.3**] of the individual units (e.g., packet, container).

12. PRODUCT PREPARATION

12.1 Remove the smokeless tobacco product from the packs or container. Include quality control samples (when applicable).

12.2 Take an appropriate representative portion of the smokeless tobacco product according to individual laboratory practice (e.g., food analysis sampling approach may be applied).

12.3 Extract the smokeless tobacco from the smokeless tobacco product.

12.4 Combine and mix sufficient smokeless tobacco product samples to constitute about 0.5–2 g of homogeneous smokeless tobacco for each test sample.

13. PREPARATION OF THE SMOKING MACHINE
Not applicable

14. SAMPLE GENERATION
Not applicable

15. SAMPLE PREPARATION
15.1 Oven preparation

Turn on the oven, and set the temperature to 100 °C ± 1 °C. Allow the oven to equilibrate for at least 1 h before use, and ensure that the temperature stabilizes at 100 °C ± 1 °C.

15.2 Preparation of samples

15.2.1 Weigh a clean, dry evaporating dish on an analytical balance, and record the weight as W_T.

15.2.2 Remove the dish from the balance, and fill it with 1–5 g of the smokeless tobacco test sample, depending on the size.

15.2.3 Place the dish with sample on the balance, and record the weight at W_1 to 0.01-g accuracy.

15.2.4 Check that the oven temperature is at 100 °C ± 1 °C, and place the samples in the oven.

15.2.5 Close the oven door, record the start time, and leave for 3 hours ± 0.5 min.

15.2.6 Repeat steps **15.2.1**–**15.2.5** for all samples.

16. SAMPLE ANALYSIS

16.1 After 3 h ± 0.5 min, remove the evaporating dishes containing the samples from the oven, and place them in a desiccator.

16.2 Allow the samples to cool to room temperature in the desiccator for approximately 30 min.

16.3 Weigh the samples in the evaporating dish on the balance. Record the weight as W_2 to 0.01-g accuracy.

16.4 Repeat step **16.3** for each sample.

17. DATA ANALYSIS AND CALCULATIONS

17.1 Calculate the percentage moisture content with the following formula:

$$M\ (\%) = \frac{W_1 - W_2}{W_1 - W_T} \times 100$$

where:

M = moisture content

W_1 = initial weight of smokeless tobacco sample and evaporating dish

W_2 = weight of dried smokeless tobacco sample and evaporating dish

W_T = tare weight of evaporating dish.

The calculated moisture content can be used to convert the concentration of an analyte presented on an as-is or wet weight basis to a dry-weight basis using the following formula:

$$C_{Dry} = C_{Wet} \times \frac{100}{100 - M}$$

where:

M = moisture content (%)

C_{Dry} = concentration of analyte on a dry-weight basis

C_{Wet} = concentration of analyte on an as-is or wet-weight basis.

18. SPECIAL PRECAUTIONS
Not applicable

19. DATA REPORTING
19.1 Report individual measurements for each sample evaluated.

19.2 Report results as specified in the overall project specifications.

19.3 For more information, see WHO TobLabNet SOP 02 [**2.4**].

20. QUALITY CONTROL

20.1 Control parameters

Note: If the control measurements are outside the tolerance limits of the expected values, appropriate investigation and action must be taken.

Note: Additional laboratory quality assurance procedures should be carried out if necessary, in order to comply with the policies of individual laboratories.

20.2 Quality control sample

To verify the consistency of the entire analytical process, analyse a reference tobacco product, such as CORESTA Reference Products (CRPs), when available, in accordance with the practices of individual laboratories.

21. METHOD PERFORMANCE SPECIFICATIONS

21.1 Limit of reporting

The limit of reporting is set to the lowest moisture content to be determined as percentage of weight.

21.2 Internal quality control

Recovery of reference material is a surrogate measure of accuracy. Recovery is determined by measuring the level of moisture in reference smokeless tobacco products. The recovery is calculated from the following equation:

$$\text{Recovery (\%)} = 100 \times (\text{analytical result} / \text{certified amount})$$

Table 1. Mean and recovery of moisture content in smokeless tobacco products

Smokeless tobacco sample	Certified value (%)	Mean moisture content (%)	Recovery (%)
CRP 1	53.7	53.87	100.3
CRP 2	53.0	51.20	96.6
CRP 3	8.1	7.67	94.7
CRP 4	23.0	24.07	104.7

22. REPEATABILITY AND REPRODUCIBILITY

An international collaborative study conducted between September 2020 and March 2021, involving 13 laboratories and four CRP smokeless tobacco products, performed according to WHO TobLabNet Method Validation Protocol and this SOP, gave the following values for this method.

The test results were analysed statistically in accordance with ISO 5725-1 [**2.5**] and ISO 5725-2 [**2.6**] to give the precision data shown in Table 2.

Table 2. Precision limits for determination of moisture content (%) in smokeless tobacco products

Reference tobacco product	n	Mean	Repeatability limit (r)	Reproducibility limit (R)
CRP1.1	9	53.87	1.02	1.93
CRP2.1	10	51.20	0.76	2.73
CRP3.1	10	7.67	0.81	2.68
CRP4.1	12	24.07	1.82	4.78

WHO TobLabNet
Official Method
SOP 14

Standard operating procedure for determination of the pH of smokeless tobacco products

Method:	Determination of the pH of smokeless tobacco products
Analytes:	pH
Matrix:	Smokeless tobacco products
Last update:	December 2021

FOREWORD

Smokeless tobacco products are gradually attracting the interest of public health organizations. A request was made by the WHO Framework Convention on Tobacco Control (WHO FCTC) Conference of the Parties (COP) at its fifth session (Seoul, 2012) to identify options to regulate chemicals in smokeless tobacco products. This document is prepared in response to the request made by the COP at its seventh session (Delhi, 2016) to the WHO FCTC Secretariat to invite WHO to finalize the standard operating procedures (SOPs) for measuring nicotine and tobacco-specific nitrosamines as requested by decision FCTC/COP6(12) 2b.ii. In pursuance of this request, WHO organized a collaborative study involving its Tobacco Laboratory Network (TobLabNet) testing laboratories, which tested materials for which there was some chemical characterization, represented a range of common forms of smokeless tobacco products and differed in physical and chemical properties. The assessment of applicability and adaptability of validated WHO SOPs to smokeless tobacco products and the recommended methods are presented in this SOP.

This document was prepared by members of the WHO TobLabNet as an analytical method SOP for measuring the pH of smokeless tobacco products. pH is one of the key influences on the nicotine delivery capacity of a product.

INTRODUCTION

In order to establish comparable measurements for testing tobacco products globally, consensus methods are required for measuring specific parameters of smokeless tobacco products. The WHO TobLabNet reviewed commonly used procedures for determining the pH of smokeless tobacco products in order to prepare a procedure as a WHO TobLabNet SOP.

This SOP was adapted from CORESTA Recommended Method No. 69 [**2.1**] to describe the procedure for determination of the pH of smokeless tobacco products.

1.   SCOPE

This SOP gives general guidelines for measurement of the pH of smokeless tobacco products by water extraction followed by pH meter readings. This method is suitable for measuring pH in the range 4–14 for smokeless tobacco products.

2.   REFERENCES

2.1   CORESTA Recommended Method No. 69 Determination of pH of Tobacco and Tobacco Products

2.2   ISO 3696:1987 Water for analytical laboratory use – Specification and test methods.

2.3 United Nations Office on Drugs and Crime. Guidelines on representative drug sampling. Vienna, Laboratory and Scientific Section, 2009 (http://www.unodc.org/documents/scientific/Drug_Sampling.pdf).

2.4 World Health Organization. Standard operating procedure for validation of analytical methods of tobacco product contents and emissions. Geneva, Tobacco Laboratory Network, 2017 (WHO TobLabNet SOP 02).

2.5 ISO 5725-1. Accuracy (trueness and precision) of measurement methods and results – Part 1: General principles and definitions.

2.6 ISO 5725-2: Accuracy (trueness and precision) or measurement methods and results – Part 2: Basic method for the determination of repeatability and reproducibility of a standard measurement method.

3 TERMS AND DEFINITIONS

3.1 *pH*: pH in smokeless tobacco products

3.2 *Smokeless tobacco*: Tobacco-containing part of a smokeless tobacco product.

3.3 *Smokeless tobacco products*: Products entirely or partly made of leaf tobacco as the raw material that are manufactured to be used by sucking, chewing or snuffing (Article 1(f) of the WHO FCTC), including snus (dry and wet), chewing tobacco or a mixture of material originating from tobacco plants.

3.4 *Laboratory sample*: Sample intended for testing in a laboratory, consisting of a single type of product delivered to the laboratory at one time or within a specified period.

3.5 *Test sample*: Product to be tested, taken at random from the laboratory sample. The number of products taken shall be representative of the laboratory sample.

3.6 *Test portion:* Random portion from the test sample to be used for a single determination. The number of products taken shall be representative of the test sample.

4. METHOD SUMMARY

Aqueous extracts of smokeless tobacco product samples are prepared, and their pH measured with a pH electrode.

5. SAFETY AND ENVIRONMENTAL PRECAUTIONS

5.1 Take routine safety and environmental precautions, as in any chemical laboratory activity.

5.2 The testing and evaluation of certain products with this test method may require the use of materials or equipment that could be hazardous or harmful to the environment; this document does not purport to address all the safety aspects associated with its use. All persons using this method have the responsibility to consult with the appropriate authorities and to establish health and safety practices as well as environmental precautions in conjunction with any existing applicable regulatory requirements prior to its use.

5.3 Special care should be taken to avoid inhalation or dermal exposure to harmful chemicals. Use a chemical fume hood, and wear an appropriate laboratory coat, gloves and safety goggles when preparing or handling undiluted materials, standard solutions, extraction solutions or collected samples.

6. APPARATUS AND EQUIPMENT
Usual laboratory apparatus, in particular:

6.1 pH meter.

6.2 Shaker (linear type) configured to hold the extraction vessels in position.

6.3 Extraction vessels: Erlenmeyer flasks (100-mL) with stoppers, 100-mL Pyrex bottles with crimp seals and septa, 100-mL culture tubes with Teflon-lined caps or other suitable flasks.

7. REAGENTS AND SUPPLIES
All reagents shall be of at least analytical reagent grade unless otherwise noted. When possible, reagents are identified by their Chemical Abstract Service (CAS) registry numbers.

7.1 Water, compliant with grade 2 of ISO 3696:1987 [**2.2**], or better.

7.2 Standard pH buffer solutions (pH = 4.01, 9.21 or > 10 according to the availability of pH calibration solutions in the laboratory).

8. PREPARATION OF GLASSWARE
8.1 Clean and dry glassware in a manner to avoid contamination.

9. PREPARATION OF SOLUTIONS
Not applicable

10. PREPARATION OF STANDARDS
Not applicable

11. SAMPLING

11.1 Sample smokeless tobacco products according to the laboratory sampling procedure. Alternative approaches may be used to obtain a representative laboratory sample in accordance with individual laboratory practice or when required by specific regulation or availability of samples.

11.2 Constitution of test sample

11.2.1 Divide the laboratory sample into separate units (e.g., packet, container), if applicable.

11.2.2 Take an equal amount of products for each test sample from at least $\sqrt{n}$[2.3] of the individual units (e.g., packet, container).

12. PRODUCT PREPARATION

12.1 Remove the smokeless tobacco product from the packs or container. Include quality control samples (when applicable).

12.2 Take an appropriate, representative portion of the smokeless tobacco product according to individual laboratory practice (e.g., food analysis sampling approach may be applied).

12.3 Extract the smokeless tobacco from the smokeless tobacco product.

12.4 Combine and mix sufficient smokeless tobacco product samples to constitute about 2–5 g of homogeneous smokeless tobacco for each test sample.

13. PREPARATION OF THE SMOKING MACHINE
Not applicable

14. SAMPLE GENERATION
Not applicable

15. SAMPLE PREPARATION

15.1 Weigh at least 1.0 g ± 0.1 g of smokeless tobacco sample into a 100-mL Erlenmeyer flask. A sample weight range of 1–2 g is suggested for this method.

15.2 Add 20.0 mL ± 0.5 mL of water into the flask, and shake gently for about 30 min.

16. SAMPLE ANALYSIS

16.1 Calibration of pH meter

16.1.1 Calibrate the pH electrode with at least two pH buffer solutions (pH = 4.01 and 9.21 or above); produce a two-point calibration within the calibration range. Calibration and measurement of samples are performed consecutively, as the two must be completed at the same temperature.

16.1.2 The electrode slope should be within 95–105% of the calculated value before the electrode can be used for sample measurement.

16.1.3 Rinse the electrode with distilled water before and after each measurement.

16.2 pH measurements of samples

16.2.1 Measure the pH of the smokeless tobacco samples within 60 min of stopping shaking.

16.2.2 After shaking, let the sample flask stand for another 20 min to allow the solution to settle, filter through Whatman paper No 41, or centrifuge at medium speed at 10 000 rpm if necessary.

16.2.3 Record the temperature. All pH measurements should be performed at room temperature, 20–25 °C ± 1° C.

16.2.4 Measure the pH to two decimal places.

16.2.5 The pH of a sample must be measured repeatedly at 15-min intervals. The average measurement for 1 h (four readings) is reported. The maximum tolerated variation in pH between readings is 0.3 units.

16.2.6 Rinse the electrode with distilled water before and after each measurement.

17. DATA ANALYSIS AND CALCULATIONS
Not applicable

18. SPECIAL PRECAUTIONS
Not applicable

19. DATA REPORTING
19.1 The test report should provide the pH results to two decimal places

19.2 For more information, see WHO TobLabNet SOP 02 [**2.4**].

20. QUALITY CONTROL

20.1 Control parameters

Note: If the control measurements are outside the tolerance limits of the expected values, appropriate investigation and action must be taken.

Note: Additional laboratory quality assurance procedures should be carried out if necessary, in order to comply with the policies of individual laboratories.

20.2 Quality control sample

To verify the consistency of the entire analytical process, analyse a reference tobacco product, such as CORESTA Reference Products (CRPs), when available, in accordance with the practices of individual laboratories.

21. METHOD PERFORMANCE SPECIFICATIONS

Recovery of reference material is a surrogate measure of accuracy. Recovery is determined by measuring the level of moisture in reference smokeless tobacco products. The recovery is calculated from the following equation.

Recovery (%) = 100 × (analytical result / certified value)

Table 1. Mean and recovery of pH in smokeless tobacco products

Smokeless tobacco sample	Certified value	Mean pH	Recovery (%)
CRP 1	8.4	7.68	92.8
CRP 2	7.7	8.05	103.6
CRP 3	7.1	7.04	99.5
CRP 4	6.2	5.95	95.7

22. REPEATABILITY AND REPRODUCIBILITY

An international collaborative study conducted between September 2020 and March 2021, involving 13 laboratories and four samples (four CRP smokeless tobacco products), performed according to WHO TobLabNet Method Validation Protocol and this SOP, gave the following values for this method.

The test results were analysed statistically in accordance with ISO 5725-1 [**2.5**] and ISO 5725-2 [**2.6**] to give the precision data shown in Table 2.

Table 2. Precision limits for determination of pH in smokeless tobacco products

Reference tobacco product	N	Mean	Repeatability limit (r)	Reproducibility limit (R)
CRP1.1	13	7.68	0.17	1.70
CRP2.1	13	8.05	0.20	1.31
CRP3.1	13	7.04	0.10	2.84
CRP4.1	13	5.95	0.10	1.00

WHO TobLabNet
Official Method
SOP 15

Standard operating procedure for determination of nicotine, glycerol and propylene glycol content in the tobacco of heated tobacco products

Method:	Determination of nicotine, glycerol and propylene glycol content in the tobacco of heated tobacco products (HTPs)
Analytes:	Nicotine (3-[(2S)-1-methylpyrrolidin-2-yl]pyridine) (CAS # 54-11-5) Glycerol (propane-1,2,3-triol) (CAS # 56-81-5) Propylene glycol (propane-1,2-diol) (CAS # 57-55-6) Triacetin (glyceryl triacetate) (CAS # 102-76-1)
Matrix:	Tobacco
Last update:	25 August 2023

FOREWORD

This document was prepared by the No Tobacco Unit of the Health Promotion Department of the World Health Organization and members of the WHO Tobacco Laboratory Network (TobLabNet), as an analytical method standard operating procedure (SOP) for measuring nicotine, glycerol and propylene glycol content in the tobacco of heated tobacco products (HTPs). The method is also applicable for the quantification of triacetin upon proper verification in a laboratory, paying particular attention to the recommended quality control criteria.

INTRODUCTION

To establish comparable measurements for testing heated tobacco products (HTPs) globally, consensus methods are required for measuring the nicotine, glycerol and propylene glycol content in the tobacco used in HTPs. WHO TobLabNet reviewed commonly used procedures for the determination of nicotine, glycerol, and propylene glycol in tobacco of heated tobacco products in order to prepare a procedure as a WHO TobLabNet SOP.

The Conference of the Parties (COP) to the WHO Framework Convention on Tobacco Control (WHO FCTC) at its eighth session (Geneva, Switzerland, 1–6 October 2018) requested the Convention Secretariat to invite WHO and WHO Tobacco Laboratory Network to: "(c) to assess whether the available standard operating procedures for contents and emissions are applicable or adaptable to heated tobacco products; (d) to advise, as appropriate, on suitable methods to measure the contents and emissions of these products"; as outlined in paragraph 2 of decision FCTC/COP8(22) on Novel and emerging tobacco products.

This SOP that was prepared to describe the procedure for the determination of nicotine, glycerol and propylene glycol content in the tobacco of HTPs, was adapted based on WHO TobLabNet SOP 11 **[2.1]**, WHO TobLabNet SOP 06 **[2.2]** and the publication by Chen et. al **[2.3]**.

1.   SCOPE

This method is suitable for the quantitative determination of nicotine, glycerol, and propylene glycol content in the tobacco of HTPs by gas chromatography (GC). This method can also be used to quantitatively determine triacetin in the tobacco of HTPs by GC, however triacetin was not validated by a collaborative trial due to the limited number of reported results. The working range of the method for nicotine is up to 50 mg/g, for glycerol, up to 500 mg/g; for propylene glycol, up to 100 mg/g and for triacetin, up to 100 mg/g (optional).

2. REFERENCES

2.1 World Health Organization. 2021. Standard operating procedure for determination of nicotine, glycerol and propylene glycol in e-liquids, Geneva. Tobacco Laboratory Network, WHO TobLabNet SOP 11.

2.2 World Health Organization. 2016. Standard operating procedure for determination of humectants in cigarette tobacco filler, Geneva. Tobacco Laboratory Network, WHO TobLabNet SOP 06.

2.3 Chen A X, Akmam Morsed F and Cheah, N P. 2021. A Simple Method to Simultaneously Determine the Level of Nicotine, Glycerol, Propylene Glycol, and Triacetin in Heated Tobacco Products by Gas Chromatography Flame Ionization Detection. Journal of AOAC INTERNATIONAL.

2.4 ISO 13276: Tobacco and tobacco products — Determination of nicotine purity — Gravimetric method using tungstosilicic acid, 2021.

3 TERMS AND DEFINITIONS

3.1 *Nicotine, glycerol, propylene glycol and triacetin (optional) content:* Individual amounts of nicotine, glycerol, propylene glycol (and triacetin) in HTP tobacco, expressed as mg/g of HTP tobacco.

3.2 *Heated tobacco products (HTPs):* A product containing tobacco or a tobacco substrate that is designed to be heated by a separate source (e.g., electrical, aerosol, carbon, etc.) to produce a nicotine-containing aerosol, which is then inhaled by users.

3.3 *Laboratory sample*: Sample intended for testing in a laboratory, consisting of a single type of product delivered to the laboratory at one time or within a specified period.

3.4 *Test sample*: Sample of HTPs, taken at random from the laboratory sample. The test sample shall be representative of the laboratory sample.

3.5 *Test portion:* Random portion from the test sample to be used for a single determination. (The amount or volume of the test sample taken for analysis, usually of known weight or volume, IUPAC definition).

Note: The number of test portions analyzed per test sample shall be adapted to sample inhomogeneity.

4. METHOD SUMMARY

4.1 Nicotine, glycerol and propylene glycol are extracted from HTP tobacco filler with a diluent consisting of 70% methanol/30% acetonitrile [7.7] [7.8] and internal standards [7.9] [7.10]. The extracts are measured by gas chromatography with flame ionization detection (GC-FID).

4.2 The ratios of the peak areas of analytes (nicotine, glycerol and propylene glycol) and corresponding internal standards, derived from the measurement of standard solutions with known concentrations, are plotted against the analyte concentration. Calibration curves used to determine the analyte content of each test portion are created by linear regression.

5. SAFETY AND ENVIRONMENTAL PRECAUTIONS

5.1 Follow routine safety and environmental precautions, as in any chemical laboratory activity.

5.2 The testing and evaluation of certain products with this test method may require the use of materials or equipment that could be hazardous or harmful to the environment; this document does not purport to address all the safety aspects associated with its use. All persons using this method have the responsibility to consult the appropriate authorities and to establish health and safety practices as well as environmental precautions, in conjunction with any applicable regulatory requirements, prior to its use.

5.3 Special care should be taken to avoid inhalation or dermal exposure to harmful chemicals. Use a chemical fume hood, and wear an appropriate laboratory coat, gloves and safety goggles when preparing or handling undiluted materials, standard solutions, diluent solutions or collected samples.

6. APPARATUS AND EQUIPMENT
Usual laboratory apparatus, in particular:

6.1 Sonicator configured to hold the vessels in position.

6.2 Gas chromatograph equipped with a flame ionization detector (GC-FID).

6.3 Capillary GC column capable of distinct separation of solvent peaks, the peaks for the internal standard, nicotine, and other tobacco components, e.g., Agilent DB-ALC1 (30 m x 0.32 mm, 1.8 μm).

6.4 Calibrated analytical balance with a readability of 0.0001 g.

6.5 Bulb pipette and syringe pipette, e.g., 1.0 ml to 20.0 ml suitable capacity for sample and standard preparation).

6.6 Class A Volumetric flask (suitable capacity for standard preparation).

6.7 Standard food grinder.

7. REAGENTS AND SUPPLIES
All reagents shall be of at least analytical reagent grade unless otherwise noted. When possible, reagents are identified by their Chemical Abstract Service (CAS) registry numbers.

7.1 Carrier gas: Helium [CAS number: 7440-59-7] of high purity (> 99.999%). Hydrogen [CAS number: 1333-74-0] of high purity (> 99.999%) can be used as an alternative carrier gas.

7.2   Auxiliary gases: Air and hydrogen [CAS number: 1333-74-0] of high purity (> 99.999%) for the flame ionization detector.

7.3   Nicotine (3-[(2S)-1-methylpyrrolidin-2-yl]pyridine) [CAS number: 54-11-5] of known purity not less than 98% **[2.4]**. Nicotine salicylate [CAS number: 29790-52-1] of known purity not less than 98% may be used alternatively.

7.4   Glycerol (Propane-1,2,3-triol) [CAS number: 56-81-5] of known purity not less than 98%.

7.5   Propylene glycol (propane-1,2-diol) [CAS number: 57-55-6] of known purity not less than 98%.

7.6   Triacetin (glyceryl triacetate) [CAS number: 102-76-1] of known purity not less than 98% (optional).

7.7   Methanol, chromatographic purity [CAS number: 67-56-1].

7.8   Acetonitrile, chromatographic purity [CAS number: 75-05-8].

7.9   Internal standard for nicotine and triacetin (optional): *n*-heptadecane (purity > 98% of mass fraction) [CAS number: 629-78-7].

7.10 Internal standard for glycerol and propylene glycol: 1,3-butanediol (purity > 99% of mass fraction) [CAS number: 107-88-0].

8. PREPARATION OF GLASSWARE
8.1   Clean and dry glassware in a manner to avoid contamination.

9. PREPARATION OF SOLUTIONS
9.1   **Diluent solution**

The diluent solution consists of 70% methanol/30% acetonitrile [7.7] [7.8] (700 mL methanol plus 300 mL acetonitrile) containing appropriate amounts of internal standards:

Pipette 0.50 mL of *n*-heptadecane [7.9] plus 2.00 mL 1,3-butanediol [7.10] into a 1-litre volumetric flask.

Dilute to volume (1 litre) with 70% methanol/30% acetonitrile [7.7] [7.8], mix thoroughly and transfer the solution into a storage container equipped with features to prevent contamination.

Note: The concentration and/or type of internal standard may be adjusted, keeping in mind the possible effect of internal standards on the sensitivity and selectivity, as well as the linear range of the method.

10. PREPARATION OF STANDARDS
Preparation of the standard solutions as described below is for reference purposes. The preparation of the standard solutions can be adjusted, if necessary.

Solvent and solutions stored at low temperatures shall be allowed to equilibrate to (22 ± 5) °C before use.

10.1 Nicotine standard stock solution (5 g/L)

Weigh approximately 500 mg of nicotine [7.3] (or 925 mg nicotine salicylate) to the nearest 0.1 mg into a 100 mL volumetric flask and dilute to volume with the diluent solution [9.1].

Mix thoroughly and store between 0°C and 4°C protected from light.

10.2 Glycerol standard stock solution (50 g/L)

Weigh approximately 5000 mg of glycerol [7.4] to the nearest 0.1 mg into a 100 mL volumetric flask and dilute to volume with the diluent solution [9.1].

Mix thoroughly and store between 0°C and 4°C protected from light.

10.3 Propylene glycol standard stock solution (5 g/L)

Weigh approximately 500 mg of propylene glycol [7.5] to the nearest 0.1 mg into a 100 mL volumetric flask and dilute to volume with the diluent solution [9.1].

Mix thoroughly and store between 0°C and 4°C protected from light.

10.4 Triacetin standard stock solution (20 g/L) (Optional)

Weigh approximately 2000 mg of triacetin [7.6] to the nearest 0.1 mg into a 100 mL volumetric flask and dilute to volume with the diluent solution [9.1].

Mix thoroughly and store between 0°C and 4°C protected from light.

10.5 Working standard solutions

10.5.1 Pipette the designated amount of nicotine stock standard solution prepared in **10.1** for the specific standard solution into 100 mL volumetric flasks, as described in Table 1.

10.5.2 Pipette the designated amount of glycerol stock standard solution prepared in **10.2** into the same set of 100 mL volumetric flasks [**10.5.1**], as specified in Table 2.

10.5.3 Pipette the designated amount of propylene glycol stock standard solution prepared in **10.3** into the same set of 100 mL volumetric flasks [**10.5.1**], as specified in Table 3.

(Optional) Pipette the designated amount of triacetin stock standard solution prepared in 10.4 into the same set of 100 mL volumetric flasks [**10.5.1**], as specified in Table 4.

10.5.4 Fill the volumetric flasks to the mark (100 mL) with diluent solution [**9.1**].

10.5.5 Store the standard solutions, protected from light, at 4–8°C.

10.5.6 Determine the final nicotine, glycerol, propylene glycol and triacetin (optional) concentrations in the standard solutions from the following equation:

Final concentration (mg/mL) = $\dfrac{x*y}{10000}$

where:

x is the original weight (in mg) of the component as weighed in **10.1**, **10.2**, **10.3** or **10.4**; and

y is the volume of the stock standard solution (in mL) as pipetted in **10.5.1**, **10.5.2**, **10.5.3** and **10.5.4**.

The final concentrations in the nicotine standard solutions are shown in Table 1, the glycerol concentrations are shown in Table 2, the propylene glycol concentrations are shown in Table 3, and the triacetin concentrations are shown in Table 4 (optional).

Table 1. Concentrations of nicotine in standard solutions

Standard	Volume of nicotine stock standard solution (**10.1**) (mL)	Total volume (mL)	Nominal nicotine concentration in final mixed standard solution (mg/mL)
1	2.0	100	0.1
2	4.0	100	0.2
3	8.0	100	0.4
4	16.0	100	0.8
5	20.0	100	1.0

Table 2. Concentrations of glycerol in standard solutions

Standard	Volume of glycerol stock standard solution (**10.2**) (mL)	Total volume (mL)	Nominal glycerol concentration in final mixed standard solution (mg/mL)
1	1.0	100	0.5
2	4.0	100	2.0
3	8.0	100	4.0
4	16.0	100	8.0
5	20.0	100	10.0

Table 3. Concentrations of propylene glycol in standard solutions

Standard	Volume of propylene glycol stock standard solution (**10.3**) (mL)	Total volume (mL)	Nominal propylene glycol concentration in final mixed standard solution (mg/mL)
1	0.6	100	0.0
2	4.0	100	0.2
3	10.0	100	0.5
4	20.0	100	1.0
5	40.0	100	2.0

Table 4. Concentrations of triacetin in standard solutions (optional)

Standard	Volume of triacetin standard solution (**10.4**) (mL)	Total volume (mL)	Nominal triacetin concentration in final mixed standard solution (mg/mL)
1	0.5	100	0.1
2	1.5	100	0.3
3	2.5	100	0.5
4	5.0	100	1.0
5	10.0	100	2.0

The range of the standard solutions may be adjusted, depending on the equipment used and the samples to be tested, keeping in mind the possible effect on the working range of the method.

All solvents and solutions must be adjusted to room temperature (22 ± 5°C) before use.

11. SAMPLING
11.1 Sample collection

Sample HTPs to obtain a representative sample, as required by applicable regulation or availability of samples.

11.2 Constitution of test sample

Divide the laboratory sample into separate sales units, if applicable.

Take a representative number of HTP sticks from at least $\sqrt{n}$ [**2.5**] of the individual sales units (e.g., packet, container) for the preparation of the test sample.

12. PRODUCT PREPARATION
12.1 Remove the tobacco from at least one sales unit of the HTP sticks from which the test portion and quality control samples (when applicable) will be formed.

12.2 Combine and homogenize the tobacco of the HTP sticks from **12.1** to constitute at least 2 g.

13. PREPARATION OF THE SMOKING MACHINE
Not applicable.

14. SAMPLE GENERATION
Not applicable.

15. SAMPLE PREPARATION
15.1 Mix and grind the tobacco from **12.2** until it is well homogenized.

15.2 Weigh about 0.2–0.3 g of the well-mixed, ground tobacco sample into a glass extraction vessel.

15.3 Add 10 mL of diluent to the sample.

15.4 Sonicate the vessel for 60 minutes.

15.5 Transfer an aliquot of the sample extract into an autosampler vial, preventing solid particles from entering the vial.

15.6 If the sample is to be stored before analysis, keep it protected from light at 4–8 °C.

16. SAMPLE ANALYSIS

GC with flame ionization detection is used to quantify nicotine, glycerol, propylene glycol and triacetin (optional) in HTPs. The analytes are separated from other potential interference on the column used. Individual analyte concentrations are derived by comparison of the peak area ratios of the unknowns with the peak area ratios of the known standard concentrations.

16.1 GC operating conditions. Example:

GC column: Agilent DB-ALC1 (30 m x 0.32 mm, 1.8 µm), or equivalent
GC parameters:
    Oven: 140 °C for 5 mins,
                                       140–180 °C at 40 °C/min,
                                       180 °C for 4 mins,
                                       180–230 °C at 5 °C/min
    Injection temperature: 225 °C
    Detector temperature: 260 °C
    Carrier gas: Helium at a flow rate of 1.5 ml/min
    Injection volume: 1.0 µl
    Injection mode: Split

Note: Adjustment of the operating parameters may be required, depending on the instrument and column conditions as well as the resolution of chromatographic peaks.

Under the above conditions, the expected total analysis time will be about 20 minutes. (The analysis time may be extended to optimize separation).

For informative purposes, Appendix 2 provides an example of GC-mass spectrometry (GC-MS) settings to be used if GC-MS is used as an alternative to GC-FID.

16.2 Expected retention times

The elution order and retention times must be verified before test sample analysis is begun.

Note: For the conditions described in **16.1**, the expected sequence of elution will be propylene glycol, 1,3-butanediol, glycerol, nicotine, triacetin and *n*-heptadecane.

Differences in for example temperature, gas flow rate and age of the column may alter retention times.

16.3 Determination of nicotine, glycerol, propylene glycol and triacetin (optional)

The quality assurance policies and procedures of the specific laboratory will determine the practices and sequence of samples analyzed. This section illustrates an example of practices and sequence of samples analyzed for determining nicotine, glycerol, propylene glycol and triacetin (optional) content in the tobacco of HTPs.

Inject aliquots of the standard solutions and sample extracts under identical conditions.

16.3.1 Condition the system just before use by injecting two 1.0-µl aliquots of a sample solution as a primer.

16.3.2 Inject 1.0 µL diluent solution [9.1] and a test standard solution under the same conditions as the samples to verify the performance of the GC system and absence of contamination of reagents used.

16.3.3 Inject an aliquot of each of the combined nicotine, glycerol, propylene glycol and triacetin (optional) standard solutions into the GC.

16.3.4 Assess the retention times and responses (area counts) of the analytes in the standard solutions. If the retention times are similar (± 0.2 min) to the retention times in previous injections, and the responses are within 20% of typical responses in previous injections, the system is ready to perform the analysis. If the responses are outside specifications, seek corrective action according to your laboratory policy.

16.3.5 Record the peak areas of nicotine, glycerol, propylene glycol and triacetin (optional) and the internal standard compounds.

16.3.6 Calculate the peak area ratios (RF) of the analyte peaks to the internal standard peak for each of the analytes in each standard solution, including the solvent blanks according to the following equation.

$$RF = A_{analyte} / A_{IS}$$

where:

RF is the peak area ratio;

$A_{analyte}$ is the peak area of the analyte peak; and

A_{IS} is the peak area of the internal standard peak.

16.3.7 Plot the concentrations of the analytes in the standard solutions (X axis) against the peak area ratios (RFs) (Y axis) as calculated in **16.3.6**.

Note: The calibration functions are expected to be linear over the specified concentration ranges.

16.3.8 Calculate the calibration function ($Y = a + bx$) by linear regression from this data and use both the slope (b) and the intercept (a) of the linear regression for calculation of analytical results. If the coefficient of determination (R^2) is less than 0.99, the calibration should be repeated. Check for individual outliers according laboratory procedures.

16.3.9 Inject 1.0 µl of each of the test portion extracts (**15.5**) and if applicable of the quality control samples [**20.3**] and determine the peak areas with the appropriate instrument software.

Note: See Appendix 1 for representative chromatograms

17. DATA ANALYSIS AND CALCULATIONS

17.1 For each test portion, calculate the ratio (Y_t) of the analyte peak areas to the internal standard peak area.

Note: The analyte peak area ratios (Y_t) obtained for all test portions must fall within the working range of the calibration curve; otherwise, standard solutions or test portions concentrations should be adjusted as necessary

17.2 Calculate the component concentration in mg/mL for each test portion according to the following equation, using the coefficients of the calibration curves determined in **16.3.8**.

$$M_t = (Y_t - a) / b$$

where:

M_t is the concentration of the analyte in the test solution in mg/mL;

Y_t is the ratio of the peak area of the analyte to the peak area of the internal standard;

a is the intercept of the calibration curve obtained by linear regression in **16.3.8**; and

b is the slope of the calibration curve obtained by linear regression in **16.3.8**.

17.3 Calculate the analyte content (m_c) in the tobacco test sample expressed in mg/g tobacco using the following equation:

$$m_c = \frac{M_t \times V_e}{m_o}$$

where:

m_c is the content of the analyte in the test sample, in mg/g;

M_t is the concentration of the analyte in the test solution, in mg/mL;

V_e is the volume of the extraction solution used, in mL; and

m_o is the mass of test portion [**12.1**] in g.

18. SPECIAL PRECAUTIONS

18.1 After installing a new column, condition it by injecting a test sample (tobacco sample) solution under the GC conditions described in **16.1**. Injections should be repeated until the peak areas (or heights) of both the component(s) and the internal standard(s) are reproducible. This may require approximately four injections.

18.2 It is recommended to elute high-boiling-point components from the GC column after each sample set (series) by raising the column temperature to the maximum allowed isothermal temperature of the column for 30 minutes.

18.3 When the peak areas (or heights) observed for the internal standard(s) in a test portion are significantly higher than expected, it is recommended to extract an aliquot of the test portion in a solution consisting of 70% methanol/30% acetonitrile [**7.7**] [**7.8**] without internal standard. This makes it possible to determine whether any component co-elutes with the internal standard, which would bias (artificially lower) results of analysis.

19. DATA REPORTING

19.1 Report individual measurements for each sample evaluated.

19.2 Report results as specified in overall project specifications.

19.3 For more information, see World Health Organization. Standard operating procedure for validation of analytical methods of tobacco product contents and emissions. Geneva, Tobacco Laboratory Network, 2017 (WHO TobLabNet SOP 02) [**2.6**].

20. QUALITY CONTROL
20.1 Control parameters

Note: If the quality control measurement results are outside the tolerance limits of the expected values, appropriate investigation and action must be taken.

Note: Additional laboratory quality assurance procedures should be carried out in compliance with the policies of the individual laboratory.

20.2 Laboratory reagent blank

To detect potential contamination during sample preparation and analysis, include a determination of diluent solution [**9.1**]. The results should be less than the limit of detection of the specific component.

20.3 QUALITY CONTROL SAMPLE

To verify the consistency of the entire analytical process, analyze a reference or quality control HTP tobacco sample in accordance with the practices of the individual laboratory.

21. METHOD PERFORMANCE SPECIFICATIONS

21.1 Note: Laboratory is encouraged to verify the method in accordance with its quality practices. Limit of detection (LOD) and limit of quantification (LOQ)

The LOD is specified as three times the standard deviation of the mean of blank determinations, and the LOQ is specified as 10 times the standard deviation of the mean of blank determinations. LOD and LOQ are shown in Table 5 (data taken from by single laboratory validation).

Table 5. LOD and LOQ of nicotine, glycerol, propylene glycol and triacetin in HTPs.

Component	LOD (mg/g)	LOQ (mg/g)
Nicotine	0.05	0.2
Glycer	1.4	4.7
Propylene glycol	1.0	1.0
Triacetin	0.2	0.8

Note: The LOD and LOQ values listed in Table 5 are outside the working ranges of this method.

21.2 Laboratory-fortified matrix recovery

The recovery of the analyte(s) spiked into the matrix was used as a surrogate measure of accuracy. The recovery of the components was determined by single laboratory validation. Tea leaves were used for the determination of the recovery because no raw tobacco without nicotine, glycerol, propylene glycol and triacetin was available. If raw tobacco was used for the determination of the recovery, the results would have been outside the working range of the method.

The recovery by single laboratory validation was determined by adding different amounts of nicotine, glycerol, propylene glycol and triacetin standard stock solutions into 10 mL volumetric flasks containing 0.2–0.3 g of tea leaves. After homogenizing the flasks for 60 minutes using a sonicator [6.1], the recovery samples were transferred into autosampler vials. For each of the spiked samples, nicotine, glycerol, propylene glycol and triacetin were determined by analyzing the individual vials in one day. This procedure was duplicated on a second day to provide the results below. The native (not spiked) tea leaves were also analyzed. The recovery is calculated using the following equation and is summarized for nicotine in Table 6, glycerol in Table 7, propylene glycol in Table 8, and triacetin in Table 9.

$$R = 100 \times \frac{\overline{c_r - c_n}}{c}$$

where:

R is the recovery, in %;

c_r is the concentration of the analyte in the recovery sample, in mg/mL;

c_n is the concentration of the analyte in the native sample, in mg/mL; and

c is the nominal analyte concentration in the recovery sample, in mg/mL.

Table 6. Mean and recovery of nicotine by single laboratory validation

Nicotine					
Day 1			Day 2		
Spiked amount (mg/g)	Mean (mg/g)	Recovery (%)	Spiked amount (mg/g)	Mean (mg/g)	Recovery (%)
5.2	5.0	89.8	5.3	5.2	91.6
20.6	20.2	97.0	21.1	20.8	95.9
41.3	40.8	97.4	42.2	41.0	95.0

Table 7. Mean and recovery of glycerol by single laboratory validation

Glycerol					
Day 1			Day 2		
Spiked amount (mg/g)	Mean (mg/g)	Recovery (%)	Spiked amount (mg/g)	Mean (mg/g)	Recovery (%)
25.1	25.0	98.1	25.2	25.4	100.5
251.0	250.7	98.2	251.4	244.1	95.2
401.6	409.2	101.1	402.3	398.2	97.2

Table 8. Mean and recovery of propylene glycol by single laboratory validation

Propylene Glycol					
Day 1			Day 2		
Spiked amount (mg/g)	Mean (mg/g)	Recovery (%)	Spiked amount (mg/g)	Mean (mg/g)	Recovery (%)
2.5	2.5	98.8	2.6	2.6	98.6
25.3	25.1	98.0	25.5	24.8	95.5
75.8	76.1	100.2	76.6	77.4	100.9

Table 9. Mean and recovery of triacetin by single laboratory validation

Triacetin					
Day 1			Day 2		
Spiked amount (mg/g)	Mean (mg/g)	Recovery (%)	Spiked amount (mg/g)	Mean (mg/g)	Recovery (%)
5.0	4.9	90.6	5.0	5.0	98.8
50.3	49.6	93.3	50.2	50.3	98.4
150.9	153.8	99.7	150.7	151.8	100.4

21.3 Analytical selectivity

The retention time of the analyte of interest is used to verify the analytical selectivity. An established range of ratios of the responses of the analytes to those of the internal standard compounds of quality control tobacco is used to verify the selectivity of the gas chromatographic measurements for an unknown sample.

21.4 Linearity

The nicotine calibration curve established is linear over the concentration range of 0.1–1.0 mg/mL. The glycerol calibration curve established is linear over the concentration range of 0.5–10.0 mg/mL. The propylene glycol calibration curve established is linear over the concentration range of 0.03–2.0 mg/mL. The triacetin calibration curve established is linear over the concentration range of 0.1–2.0 mg/mL (optional).

21.5 Possible interference

The presence of flavourings can cause interference, due to a similar retention time to one of the analytes or internal standard compounds.

22. REPEATABILITY AND REPRODUCIBILITY

An international collaborative study conducted from January 2022 to June 2022, involving 12 laboratories and three heated tobacco product samples, performed according to WHO TobLabNet Method Validation Protocol and this SOP, gave the following values for this method.

The test results were analysed statistically in accordance with ISO 5725-1 [2.7] and ISO 5725-2 [2.8] to give the precision data shown in Tables 10–12.

Table 10. Precision limits for the determination of nicotine content (mg/g) in the tobacco (substrate) of heated tobacco products

Heated tobacco products	n	$\hat{m}$	Repeatability limit	Reproducibility limit (R)
HTP1	10	13	1	5
HTP2	11	12	0	4
HTP3	11	13	1	5

Table 11. Precision limits for the determination of glycerol content (mg/g) in the tobacco (substrate) of heated tobacco products

Heated tobacco products	n	$\hat{m}$	Repeatability limit	Reproducibility limit (R)
HTP1	11	133	16	33
HTP2	10	108	11	27
HTP3	11	187	17	52

Table 12. Precision limits for the determination of propylene glycol content (mg/g) in the tobacco (substrate) of heated tobacco products

Heated tobacco products	n	$\hat{m}$	Repeatability limit	Reproducibility limit (R)
HTP1	12	3	0	2
HTP2	11	4	0	2
HTP3	9	6	0	1

23. BIBLIOGRAPHY

23.1 United Nations Office on Drugs and Crime. 2009. Guidelines on representative drug sampling. Vienna, Laboratory and Scientific Section (http://www.unodc.org/documents/scientific/Drug_Sampling.pdf).

23.2 World Health Organization. 2017. Standard operating procedure for validation of analytical methods of tobacco product contents and emissions. Geneva, Tobacco Laboratory Network (WHO TobLabNet SOP 02).

23.3 ISO5725-1. 2019. Accuracy (trueness and precision) of measurement methods and results — Part 1: General principles and definitions.

23.4 ISO 5725-2. 2019. Accuracy (trueness and precision) or measurement methods and results — Part 2: Basic method for the determination of repeatability and reproducibility of a standard measurement method.

23.5 Report of a collaborative study for the validation of an analytical method for the determination of nicotine, glycerol and propylene glycol content in the tobacco of Heated Tobacco Products (HTPs) (in press).

ANNEX 1

Typical chromatograms obtained in the analysis of HTP for the determination of nicotine, glycerol propylene glycol and triacetin (optional) in the tobacco of HTP

Fig. A1.1. Example of a chromatogram of a standard solution with nicotine concentration of 0.4 mg/mL, glycerol concentration of 4.0 mg/mL, propylene glycol concentration of 0.5 mg/mL and triacetin concentration of 0.5 mg/mL.

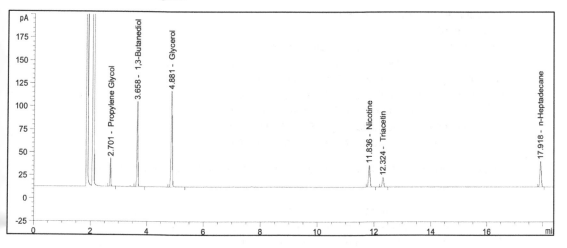

Fig. A1.2. Example of a chromatogram of an extract of HTP tobacco (substrate)

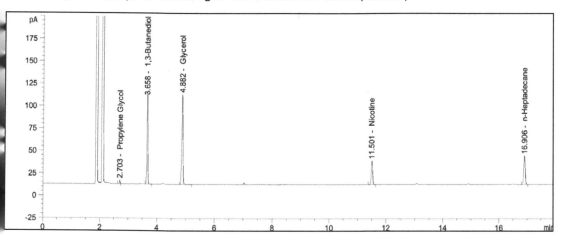

ANNEX 2
GC-MS settings for alternative measurement technique

Specific GC conditions, like carrier gas, flow rate and total run time, can be adapted to specific MS needs.

MS operating conditions:

Transfer line temperature: >= 180°C
Dwell time: 50 msec
Ionization mode: Electron Ionization (electron energy 70 eV)
Detection: propylene glycol: m/z 61 (quantifier ion) 45 (confirmation ion)
glycerol: m/z 61 (quantifier ion) 43 (confirmation ion)
nicotine: m/z 162 (quantifier ion) 133 (confirmation ion)
triacetin: m/z 145 (quantifier ion) 103 (confirmation ion)
quinaldine: m/z 143 (quantifier ion) 128 (confirmation ion)
n-heptadecane: m/z 240 (quantifier ion) 85 (confirmation ion).

Use of this internal standard is not recommended for GC-MS detection due to its low specific mass spectrum.

The recommended internal standard for GC-MS is the deuterated form of the analyte. Alternative internal standards, such as quinaldine or n-heptadecane can be used. The user should verify the performance of the alternative internal standards in own laboratory to ensure the quality control criteria is met.

GC-MS data are not included in the collaborative trial to establish the repeatability and reproducibility data of this method.